SAUCER WISDOM

BOOKS BY RUDY RUCKER

NOVELS

Realware (forthcoming)
Freeware
The Hacker and the Ants
The Hollow Earth
Wetware
The Secret of Life
Master of Space and Time
The Sex Sphere
Software
White Light
Spacetime Donuts

NONFICTION

Mind Tools
The Fourth Dimension
Infinity and the Mind
Geometry, Relativity and the Fourth Dimension
Saucer Wisdom
All the Visions

COLLECTIONS

Seek!
Transreal!
The Fifty-Seventh Franz Kafka

SAUCER WISDOM

RUDY RUCKER

FORGE®

A TOM DOHERTY ASSOCIATES BOOK
NEW YORK

Edited by David G. Hartwell

A Forge Book
Published by Tom Doherty Associates, LLC
175 Fifth Avenue
New York, NY 10010

Forge® is a registered trademark of Tom Doherty Associates, LLC.

Designed by Lisa Pifher

Library of Congress Cataloging-in-Publication Data

Rucker, Rudy v. B. (Rudy von Bitter), date
   Saucer wisdom / Rudy Rucker. —1st ed.
      p.   cm.
   "A Tom Doherty Associates book."
   ISBN 0–312–86884–7
   I. Title.
PS3568.U298S28   1999
813'.54—dc21   99–22076

First Edition: July 1999

Printed in the United States of America

0 9 8 7 6 5 4 3 2 1

For Greg Gibson, Nick Herbert, and Dick Termes

CONTENTS

Introduction by Bruce Sterling   13

CHAPTER ONE
Another Nut?   17

CHAPTER TWO
Frank Shook Time   28

CHAPTER THREE
Calling the Saucers   41

CHAPTER FOUR
Notes on the Future of Communication   54

CHAPTER FIVE
Frank and Peggy   90

CHAPTER SIX
Notes on Aliens   105

CHAPTER SEVEN
Notes on Future Biotechnology   139

CHAPTER EIGHT
The *Mondo* Party   173

CHAPTER NINE
Missing Time   188

CHAPTER TEN
Notes on Femtotechnology   207

CHAPTER ELEVEN
Notes on Transhumanity   239

CHAPTER TWELVE
The End?   276

Index   279

TABLE OF FIGURES

Figure 1: A History of Paratime 32
Figure 2: 3D Paratime Prevents Worldline Crossings 33
Figure 3: Stopping Time with Paratime 35
Figure 4: A Paratime Shortcut........................... 36
Figure 5: Hummingbird Alien............................ 38
Figure 6: Three Cameras and Three TVs.................. 49
Figure 7: The Saucer Watching the Grandchildren........... 58
Figure 8: Dragonfly and Gnat Cameras 60
Figure 9: Carlo Zaps the Dragonflies..................... 62
Figure 10: Slug-blade 67
Figure 11: Dustin the Snail Man......................... 68
Figure 12: Beanbag Chair with Live Paisley 69
Figure 13: Texas Machine Language....................... 72
Figure 14: Soft Television Set........................... 73
Figure 15: Kitchen Sluggies 75
Figure 16: Larky's Brain Show........................... 78
Figure 17: Larky and Lucy 81
Figure 18: Kotona's Flanders 87
Figure 19: Starfish Alien................................ 109
Figure 20: Flesh-Glob Aliens 111
Figure 21: Gnome Alien................................ 112
Figure 22: Turla and Dak 118
Figure 23: Butterfly People............................. 121
Figure 24: Lulu with an Alla and a Big Tongue 122
Figure 25: Starfish Aliens on a Tower Beneath a
           Black Hole Sky 124
Figure 26: Boring Worldlines and Gnarly Worldlines 126
Figure 27: Aliens Decrypting Themselves 127
Figure 28: Ladder Wetware, Sheet Wetware, and
           Block Wetware 132
Figure 29: The Shuggoth............................... 133
Figure 30: Sunspot Vortex Threads....................... 135
Figure 31: The Fractal-Line "Mind" of a Sunspot 136
Figure 32: Suzanna in the Shower with Big Tongue and
           Lickin' LuvSlug............................. 148
Figure 33: ShushBees vs. the Leaf-Blowing Neighbor 151
Figure 34: Neatnik Bird 153
Figure 35: Dona and the Compsognathii 156
Figure 36: A Rhamphorhynchus Chasing a Cat 158

Figure 37: Jose and Amparo with Knifeplants 161
Figure 38: Casas Gordas 165
Figure 39: The Bosch House 166
Figure 40: Kip Kelp Sea Homes 168
Figure 41: Nine Morphs 171
Figure 42: Dragon-Saucer Coming Out of Peggy's TV 211
Figure 43: The Rope Aliens 213
Figure 44: Rust and the Alla 215
Figure 45: A Piezoplastic Rocket Suit 223
Figure 46: Eden in a Hollow Asteroid 225
Figure 47: Copying a Space Bump by Folding 228
Figure 48: Joe's Third Thumb 230
Figure 49: Man Getting Twinned 234
Figure 50: Lulu and Her Wand 236
Figure 51: Durleen's Neck Fingers 246
Figure 52: Otis's Extra Hands 250
Figure 53: The Tub and the Eyeball 252
Figure 54: Silla the Mermaid 255
Figure 55: A Sluggie Sack Brain Injection 260
Figure 56: Boba Sneezing on His Ohmies 264
Figure 57: Launching a Space Bug 268

TIMELINE FOR THE FUTURE HISTORY OUTLINED IN FRANK SHOOK'S NOTES

2030   Piezoplastic.

2040   Lifeboxes.

2050   Limpware Engineering.

2060   Dragonflies.

2070   Sluggie processors. Soft TVs. Smart furniture.

2080   Radiotelepathy. The uvvy.

2090   The S-cube. Recording dreams. Superanimation.

2102   Glork energy currency. First attempts at personality encryption.

2110   Commercial genetic engineering. Big Tongue. ShushBees.

2115   Knifeplants. Devilberries. Casas Gordas.

2120   Phido pet construction kit. Pet dinosaurs.

2125   Grown Homes. The Bosch House. Sea Homes from kelp.

2280   Aug dog people alter their bodies.

2290   Archipelago people with disconnected hands. Mermen and mermaids.

2350   Programming clones with lifebox files.

2550   Programming ohmie clones with prion infection.

3001   The alla matter transmuter.

3002   People begin using allas to live in asteroids.

3003   Strange matter used for 3ox copying of objects and of living beings.

3150   Uvvies become internal organs.

3400   People like Ang Ou get hundreds of bodies. Spacebug people.

3666   Teleportation.

4004   People become saucerians, teleporting across the universe.

4050   People free to move in the astral plane.

INTRODUCTION

Bruce Sterling

I have been asked by Professor Rucker's publishers to explain a few facts about this extraordinary work.

On the face of it, it would seem that one of the professor's friends has been abducted by aliens in flying saucers. These friendly creatures can time-travel, so they conveyed our witness step-by-step into the deep future, and they showed him any number of amazing marvels. No, it's even better than that. They brilliantly lectured to him about everything he saw, just as if they were philosopher novelists like Olaf Stapledon. While he was a contactee, our guy took copious notes on a yellow pad. He did a lot of cool eyewitness drawings too. And it's all in here.

I have studied the manuscript closely, and I've reached an important conclusion. While the twisted super-science marvels of the future seem only too plausible to me, I pretty well know that Rucker's pulling our leg with the "friend" part. I strongly question the reality of this supposed "Frank Shook" character, the "friend" who does all the close encounters. As portrayed, he's supposed to be a guy who is more imaginative than Rudy Rucker.

As if.

This is where I must balk. I can take all the other stuff: high-speed in-line skates made of slug slime, women with human fingers growing out of their necks, time as a three-dimensional helix, prion-infected clones. I'm totally down with all that; in fact, having read it all, I just wanted more, *more*. What I can't swallow is "Frank Shook," a cyber-hippie so totally spaced that he makes Rudy Rucker seem like a normal, aw-shucks kinda guy.

You see, this telling detail, this fatal discrepancy, absolutely proves to me that this book is not, in fact, "real." If you're examining *Saucer Wisdom* imagining that Rudy (or some fictional "Frank Shook") has been actually logging a lot of onboard saucer time, well, you can knock that off right now. Rudy Rucker made up the flying saucer part. There is no actual flying saucer. The saucer is not an interplanetary faster-than-light device. It's what we professional authors like to call a *narrative device*.

I'm going to spill the beans as directly as I can here: *Saucer Wisdom* is a work of popular science speculation. It's a *nonfiction book*, in which Professor Rucker takes a few quirky grains of modern scientific fact, drops then into the colorful tide pool of his own imagination, and harvests a major swarm of abalones, jellyfish, and giant anemones. This book is not the puerile raving of a UFO-stricken madman, but a *firmly controlled, intelligent* hallucination—it's the science fictional imagination in an almost pure, virginal state. Through the simple device of declaring this book to be a flying-saucer confessional, Rucker has instantly removed the bad plotting, the ray-gun melodrama, the thumb-fingered boy-meets-girl elements, the cute Spielberg dinosaur-fodder children, and all those other things that make the "fiction" in science fiction so depressing. Instead, we're getting right down to the vital hot stuff here: raw futuristic speculation, coming straight at your brain like a fiery wheel in the sky. To create a narrative like this, it might *help* to be a demented saucer nut like "Frank Shook," but raving this well and at such a sustained creative intensity takes more mental horsepower than the Shooks of the world will ever manage to muster.

Rudy Rucker can do this sort of thing very well. Thanks to his "transrealist" compositional habits, conventional narrative structures do not trouble him (as I know only too well, having cowritten two stories wtih him in the past). What Rucker has created in this book is not "science," not "fiction," and not conventional "science fiction," but a dainty little genre all its own: science/fiction, you might call it. With the emphasis on the slash.

Pop science writers didn't use to treat "science" in this boisterous way, but there might well be a trend here; there may be a real future in this. *Saucer Wisdom* is a book by a well-qualified mathematician and computer scientist, a veteran pop science writer, in which "science" is treated, not as some distant and rarefied quest for absolute knowledge, but as naturally great source material for a really long, cool rant. The book is a brilliant series of Kerouac "sketches" and Burroughs "routines." It's Beat pop science. *Saucer Wisdom* has a genuine beatnik sensibility: there's lowlife, plenty of road travel, dope and booze problems, spells of petty thievery, and romantically fulfilling moments where you get a decent Mexican soup in a cool roadside café, and life suddenly seems full of beatitude and sweet illumination.

Without being in any way New Agey or mush, *Saucer Wisdom* is a very spiritual book. It's the kind of book that would probably do actual saucer nuts a world of good, were they to pluck it off the shelf in the midst of their dopey quest for ancient astronauts or a desperate spiral down toward the poisoned pudding of a cult suicide. Luckily, actual saucer devotees almost never read entire introductions, so they will have no idea what I have been actually going on about in this prefatory note. Instead, they will skim the first sentence, and then read the last paragraph, and then dive right for the cool stuff. So for their sakes, let's move right on, shall we?

I devoutly hope that all those fascinated by flying saucers will put aside the dangerous illusions that have been foisted on them by lesser adepts. They should buy this book at once, and put themselves—entirely and trustingly—in the capable hands of their friend and mine, Professor Rudolf von Bitter Rucker.

ONE

Another Nut?

I've always liked the idea of flying saucers. To me, UFOs mean fun. I think of them as a cheerful icon of instant strangeness and enjoyably cheesy science fiction.

Although there have been some excellent UFO movies and TV shows, the books on the subject tend to be emotional downers with very little intellectual content. One of my goals in publishing *Saucer Wisdom* is to make available a UFO story that is amusing and filled with new ideas.

The story is about my encounters with a man I call "Frank Shook"—and about Frank's alleged encounters with aliens. Do I believe everything that Frank told me? It doesn't matter. What counts is that Frank's visions make up a remarkably detailed and coherent set of speculations about humanity's future. Far more than being a work of ufology, *Saucer Wisdom* is a futuristic vision of the coming millennia. Not to mention the fact that it's a hoot.

A difficulty in presenting *Saucer Wisdom* to the public is that many people have very strong fixed ideas about UFOs.

On the one hand, serious scientists and intellectuals have little patience for ufology. The fact that I am not at all interested in debunking Frank Shook's claims could well undermine my academic respectability, such as it is. But I feel Frank's ideas are important enough for me to take this risk.

On the other hand, there are many people for whom UFOs are something much too serious to be used in light-spirited intellectual romp such as *Saucer Wisdom*. For a believer, UFOs can be dark and personal and all too real. And believers are everywhere. The most diverse kinds of people have found themselves drawn into the ongoing worldwide obsession.

There is a level at which UFOs are indeed more than a cultural joke. If nothing else, they represent something important about the human psyche. So before I begin the adventures of Frank Shook, I'd like to insert a brief section containing my opinions about the meaning of UFOs and about the deplorable state of contemporary ufology in general. But if you're eager to get on with the action, you can skip this section.

UFOLOGY

In the 1950s there was a widespread feeling that the saucers were here to bring some kind of solution, perhaps to the then-paramount problem of the cold war. As the great thinker Carl Jung wrote in 1958, "The UFOs . . . have become a living myth. We have here a golden opportunity of seeing how a legend is formed, and how in a difficult and dark time for humanity a miraculous tale grows up of an attempted intervention by extraterrestrial 'heavenly' powers. . . ."[1]

For Jung, the circular UFO is a mandala symbol, representing an integration of the individual psyche with the forces of the cosmos. The flying saucer is thus a projection of the human desire for wholeness and unity. This insight of Jung's is simple and deep. The fact is that it makes people feel good to look at images of flying saucers. There is a feeling of safety and completion in these round, hovering entities.

These positive feelings are undoubtedly connected to our very earliest life experiences. Look back to the dawn of your life, back when you were part of—or very nearly part of—your mother. Your mother's

---

[1] C. G. Jung, *Flying Saucers: A Modern Myth of Things Seen in the Skies* (Princeton: Princeton University Press, 1978), 16–17. (Originally published in 1958.)

breast is the very first "round, hovering entity" that you encounter. Your mother is the original whole of which you were a part. The common use of the phrase "mother ship" for large UFOs is no accident.

In a healthy adult, the striving for wholeness is quite different from a return to the womb. Rather than longing to regress to infancy, we try instead to become capable of being parents ourselves. By an outward expansion of knowledge and compassion we become well-rounded, we learn to encompass multitudes, and if we are lucky we become parents or teachers who nurture and foster the young. One might say that in attaining emotional maturity, we *become* a womb rather than trying to reenter it. But this biological formulation leaves something out.

At the deepest level, our ultimate parent is the universe, or the God that underlies it. In maturing, we strive to become more at one with this pervasive divinity, to grow closer to the great ground of all being. This is a quest that is inherently religious, although "religion" can mean a pure spirituality rather than the adherence to the teachings of any particular human sect. And, as with the womb, the drive is not to annihilate oneself back to zero, but rather to expand one's circle of compassion out towards the infinite. In the words of the philosopher Blaise Pascal, the cosmos is a "sphere whose boundary is nowhere and whose center is everywhere."

In this connection, there was an interesting bit in the very first episode of *The X-Files*. A poster in Mulder's office shows a picture of a flying saucer. And beneath the picture is the caption: *I Want to Believe.* If we have a deep need to believe in something whole and integrated, what better symbol than a disk in the sky?

Another aspect of the roundness of the flying saucer is that it corresponds, as Jung remarks, to the shape of a mandala. Mandalas are diagrams that people in every culture spontaneously use to represent the geography of the psyche. It is customary to organize a mandala by placing opposing forces at opposite sides of the circle. The very simplest mandala of all, the yin-yang, pairs up paisley-shaped teardrops that are variously colored black-white or red-blue. The points of the compass form a familiar four-sided or eight-sided mandala. The zodiac, or wheel of the months, makes up a twelve-sided mandala. Contained within the geometry of the mandala are three key principles: The first principle is that any force has an opposing force. The second principle of the mandala is that any one axis between a force and its opposition can be transcended by looking at some other axis. For example you can move beyond an east-west conflict by thinking about a north-south synthesis. The third teaching of the mandala is that a state of complete balance involves an awareness of all the forces along all the axes. If the celestial saucers are harbingers of wholeness, it makes perfect sense for them to be shaped like mandalas.

Jung noted that the sexual instinct and the drive for power readily tend to obscure the reality of the quest for wholeness. As he puts it:

> The most important of the fundamental instincts, the religious instinct for wholeness, plays the least conspicuous part in contemporary consciousness because . . . it can free itself only with the greatest effort . . . from contamination with the other two instincts. These can constantly appeal to common, everyday facts known to everyone, but the instinct for wholeness requires for its evidence a more highly differentiated consciousness, thoughtfulness, reflection, [and] responsibility. . . . The most convenient explanations are invariably sex and the power instinct, and reduction to these two dominants gives rationalists and materialists an ill-concealed satisfaction: they have neatly disposed of an intellectually and morally uncomfortable difficulty . . .[2]

During the 1980s and 1990s, this is exactly what happened to ufology. The notion of wholeness was supplanted by notions of sex and power, and the UFO stories became accordingly unwholesome and paranoid. On the one hand, the mythos was tainted by concepts relating to society's pervasive, icky concern with sexual molestation and the politics of reproduction. And on the other hand, much of the energy of ufologists has been diverted into infantile fears that an all-powerful government

---

[2]Ibid., 38.

has been hiding saucer contacts from us. Just as Jung warned, concepts of sexuality and power have utterly eclipsed the concepts of higher consciousness.

Let me give some quick examples of the role of sex in modern ufology. Although Whitley Strieber's *Communion: A True Story* is in some ways an interesting book, it prominently features an alien proctoscope. "The next thing I knew I was being shown an enormous and extremely ugly object . . . at least a foot long, narrow, and triangular in structure. They inserted this thing into my rectum."[3] Sadly enough, this repellent scene seems to have struck a deep cultural resonance. Many more examples of this nature are to be found in the hypnotically evoked case studies described in John Mack's *Abduction: Human Encounters with Aliens* (New York: Charles Scribner's Sons, 1994).

What are Mack's abduction scenarios like? Much of a dreary muchness. You're in bed or in a car, usually asleep. You see a light. You float up into the air and into a flying saucer. Inside the saucer a tall alien who reminds you of a doctor probes at your genitals and sticks things up your butt. If you are a man, the "doctor" masturbates you to joyless orgasm, and if you are a woman, the "doctor" extracts eggs from your ovaries. Then you wake up back in your car or in your bed. Is this pathetically infantile scenario really what one might expect superhuman aliens to do? Would godlike beings fly halfway across the galaxy simply to perform what Mack calls "urological-gynecological procedures"?

Modern ufology's obsession with political power is equally inane. Book after book appears about alleged government cover-ups. One of the reasons for the success of the 1996 film *Independence Day* is that it gave such a satisfyingly vivid depiction of what has become more and more of a core belief: the U.S. government has several intact flying saucers in its possession, as well as some alien corpses and even a few live-alien saucer crew members. Mesmerized at the thought of so vast a political conspiracy, today's ufologists engage in a never-ending discus-

---

[3]Whitley Strieber, *Communion: A True Story* (New York: William Morrow, 1987), 21.

sion of amateurishly forged ''top-secret government documents'' that supposedly describe high-level contacts with aliens. How sad that in the 1990s a close encounter is likely to be with Xeroxed pseudo-bureaucratic gobbledygook—instead of with a flaming wheel from the sky.

What has happened in contemporary ufology is that sexual fantasies and conspiracy theories have clouded over any possible higher truth in the UFO experience. Where we might have hoped to find creative ideas and enlightenment, we find only clichés and hysterical fear.

One of my aims in setting *Saucer Wisdom* before the public is to try and alter this trend. Thanks to my contacts with Frank Shook, I'm in the fortunate position of being able to present a radically new style of ufology, which includes a great deal of completely new information about aliens and future worlds. I hope you enjoy Frank's tales as much as I have.

FRANK SHOOK

I first met the remarkable Frank Shook after a public lecture I gave at the Angelico Auditorium of Dominican College in San Rafael, California. It was a damp, springlike February day in 1992. I spoke on my popular science book *The Fourth Dimension* to an audience of perhaps two hundred. Three or four people approached me after my talk for autographs, or simply to say that they liked my books. After my fans had finished with me, one person remained, a smiling man about my age and height, but much thinner. For whatever reason, I immediately pegged him as an eccentric. He had medium-length brown hair and was clean-shaven. His eyes were alert behind his black-rimmed spectacles. His tone was enthusiastic and confiding, as if we were old friends.

''I've underlined a lot of things in your book *The Fourth Dimension*,'' he began, not bothering to introduce himself. ''It's material I've had occasion to think about pretty deeply. The thing is—'' He essayed a brief, direct glance into my eyes. ''I have, um, a lot of technical information. I'm having trouble putting it into words.''

''Do you mean you want to write a book?'' I asked him. Over the years I've been approached by any number of fringe-science devotees, and they can be very persistent. It's not unusual for them to expect me to help them in getting published. Some have even asked me to help them write. My interlocutor's next sentence confirmed my expectations.

"I need to get rid of the information, get it out of me, and I thought maybe you could help me process it."

"I'm not clear on what kind of information you mean."

"Well . . . it's about some unusual experiences I've had relating to space and time. I've been encountering another order of reality."

For all I knew, this man's notion of "another order of reality" consisted of studying astrology. Or taking drugs.

"Tell me something specific," I challenged him. His smile faded and he looked uneasily around the big, empty hall, as if afraid of being overheard. But everyone else had left, except for my hosts at the other end of the auditorium, now busy turning things off. Still the man hesitated, and I began to wonder if he actually had anything to say. "While we're talking, let's head towards the exit over there," I suggested. "They have to close this place up."

We stepped down off the dais and headed down the aisle together. "I'm not sure they want me to be talking with you at all, Rudy," the man finally said. "But okay, here's a hint. What I want to tell you involves *three-dimensional time*."

"All right!" I exclaimed. I'm predisposed to like any theory about how time might be different. Linear one-dimensional time is a drearily familiar past/present/future death trap I've always longed to escape.

Seeing my interest, the man's smile returned. "I thought that would get you going!"

"So what's your name?" I asked him.

"Frank Shook."[4]

Just then somebody did something that made the auditorium public-address system begin giving off a shrill drone of feedback. The grainy

---

[4]In true fact, "Frank Shook" is not really the name he told me. After we began working together, my informant requested that I use the pseudonym in our book, and that I change the names of our hometowns. Furthermore, so as not to cause my wife unnecessary discomfort, I'm using a pseudonym for the first name of the *Saucer Wisdom* character who (inadequately) represents her.

squealing disturbed Frank Shook inordinately. He waved his hands back and forth in a frightened "all-bets-are-off" kind of gesture—then whirled and ran out through a nearby side exit. I followed him outside, only to see his lean form striding off through the dusk and the gentle rain. Apparently he'd decided that "they" didn't want him to talk with me—whoever "they" were.

None of my hosts had ever heard of any Frank Shook. For a few months I'd think of him and wonder about three-dimensional time, but then I forgot him.

I occasionally review science-related books for the *Washington Post*, and in April of 1994 I wrote a negative review of John Mack's book *Abduction: Human Encounters with Aliens*, making the same point I already mentioned above: that the UFO experience, if it has any validity at all, surely must consist of more than spooky dreams and juvenile sex fantasies. I was a little worried about publishing the review. On a rational level, I was concerned that some saucer believers might harass me. And on a deeper level, I was worried that maybe, just maybe, the aliens themselves would decide to teach me a lesson.

In the weeks after the review came out, I got a few mildly worded letters of protest from Mack adherents—but nothing else happened.

Then on Tuesday, May 31, 1994, my wife Audrey and I returned home from a four-day trip to Eugene, Oregon. Mixed in with the humdrum voice mail on our answering machine were three very curious messages, delivered by a man whose voice I didn't immediately recognize. The messages were singular enough that I took the trouble to write them down.

"You're just making fun of Mack's abductees because you're scared of aliens. You think you've got the whole world in a little science box. But the aliens are everywhere, Rudy, they're all around us. Okay, now I've done it, I've told you. I'm gonna hang up and see what happens." Click. The voice sounded both frightened and exultant.

I let the messages keep playing. A greeting from our niece. A call from the bicycle repair shop. A reminder from the dentist. And then the voice was back.

"Hi, Rudy, this is Frank Shook. I was scared to leave my name yesterday. But I talked with the aliens again last night and it *is* okay for me to tell you." A big, shaky sigh. "Rudy, why don't you come out to our cabin, say at one on Thursday, June second? I can explain everything then. Remember how I said something about three-dimensional time? This has a lot to do with it. The cabin's a little hard to find—we're up in the Santa Cruz Mountains—so you should meet me at Carlita's Mexican restaurant. Here's how you get there—'' And he proceeded to leave detailed instructions, sounding more and more calm and businesslike.

The very last message of all on our machine was from Frank Shook as well, bright and chipper.

''Rudy, this is Frank. Are you out of town? Or not answering?'' Long pause. ''I was rereading your review in the library today. I thought you'd like to know that the aliens have never fiddled with my privates. For me, this isn't about sex at all. I'm still hoping to see you Thursday. Carlita's, one o'clock!''

Audrey was quite upset over these messages. ''How did this *nut* get hold of our phone number?'' she cried. ''What if he comes after you!''

''It sounds pretty bizarre,'' I agreed. ''I have met him, though, and he didn't seem physically threatening. It was after that talk I gave in San Rafael a couple of years ago. Frank Shook. He must have seen the *Post* review, remembered that I live around here, and looked us up in the phone book.''

''Oh, why did you have to write that stupid review? What do you care about flying saucers!''

I rewound the tape and played the messages again. I still wanted to know about three-dimensional time. And I was intrigued that this man flat-out said he'd talked with aliens. ''You know, I think I *should* go over to Frank Shook's house on Thursday. That would be day after tomorrow. It might be really interesting. If I go, I can call you as soon as I get there and give you his number and have you call me back. That way he'll know that someone knows I'm there, and if I want to leave, I'll pretend you said I had to hurry home.''

''Don't do it,'' said Audrey. ''What if he wants to kill you?''

''I don't think he's that kind of guy. A little weird, sure, but he smiles a lot and—and he's full of interesting ideas.''

''The smiling ones are the worst kind!''

I listened to the messages for a third time. ''Did you notice he says '*our* cabin,' Audrey? That probably means he has a wife. It's a good sign if he has a wife.''

"How do you know that when he says '*our* cabin' he isn't talking about himself and the termites that live in his gunjy saucer-nut brain?"

A PRAYER

Tuesday night, for the first time in years, I dreamed I saw a flying saucer. It was transparent, sketched in lines of pale light against a blue sky. Its form was the traditional saucer shape of a gently curved disk with a domed central cabin.

Wednesday I went out walking in the hills with my orange-and-white collie-beagle dog Arf. It was a calm, warm spring day. Arf and I found a little meadow by some oak trees, and we lay down there in the coarse green grass. Staring up into the blue sky, I remembered my dream of seeing the saucer. But maybe the aliens didn't use saucers. "The aliens are everywhere," Frank Shook had said. "They're all around us."

Over the years I've occasionally had the feeling of seeing things out of the corner of my eye—fleeting things that rush past too fast to observe. Might there be creatures who move across time as readily as we move through space?

The shadows beneath the oaks were pierced by bright shafts of sunlight, alive with drifting motes of dust and pollen. I remembered a passage in Diogenes Laertius's third-century *Life of Pythagoras*: "The Pythagoreans also assert that the whole air is full of souls."[5]

The sky looked so big and blank—both full and empty. There was nothing between outer space and me, nothing between me and the Sun and the worlds beyond. The faint noises of the woods and meadow began to seem alien, began to stick together in new shapes. I had to fight back a spasm of fear.

Now I grew angry at myself for being afraid.

It would be madness to start being scared of being outdoors alone. Solitude and nature are precious to me. I would never want to be some

---

[5]Kenneth Sylvan Guthrie, ed., *The Pythagorean Sourcebook And Library* (Grand Rapids, Michigan: Phanes Press, 1987), 149.

timid person whose idea of hiking is walking around a mall, whose notion of the seaside is a Fisherman's Wharf shopping mall, whose concept of adventure is Disneyland, whose fling with sin is Las Vegas. Yes, I did want to pursue Frank Shook's notion of aliens all around us—but not if his strange ideas ended up frightening me into being one of mass culture's obedient sheep.

You don't have to be afraid, an inner voice seemed to say. *Trust in God!*

God? These days it was highly unusual for me to think of that topic. As the son of a minister, I'd been hustled off to church Sunday after Sunday, year after year. But from my teen years on, the rituals had struck me as empty. I felt that people went to church simply because it made them feel good to be in a crowd of like-minded individuals. And with the coming of the religious right and televangelism, my indifference to Christianity had soured into fear and contempt.

In my twenties and early thirties, I'd become interested in mysticism. I adopted the notion of a supernal yet immediate One Mind, a cosmic white light that shines through ordinary objects like sunlight through stained-glass windows. In my usual scholarly fashion, I read up on mysticism, enjoying such classics as Aldous Huxley's *Perennial Philosophy* and D. T. Suzuki's *Introduction to Zen Buddhism*. I even reached the point of mentally codifying mysticism into three concise statements: All is One, the One is Unknowable, and the One is Right Here.

But as my thirties stretched on and gave way to my forties, my theories about mysticism had come to seem like a dry, academic game. What good, after all, was some metaphysical notion of a One Mind? My mother grew ill and died; my father had a debilitating stroke; my own health became less reliable. I was haunted by one of the last things my father had said to me before his stroke crippled him: "Rudy, all I know about life is this: *you get old and you die.*"

Lying there on the hilltop with Arf, I felt the sun beating down on my closed eyelids, filling my eyes with bright light. *God is everywhere*, I thought, trying the notion on. *God can hear me.* And finally, for the first time in years, I let myself pray. *Dear God, please be with me. Protect me and let me do your will.*

TWO

Frank Shook Time

Thursday morning, June 2, 1994, I set off to meet Frank Shook. His hometown—which we'll call San Lorenzo—is tucked into the Santa Cruz Mountains at the base of the San Francisco peninsula. I took a two-lane road through the woods to get there.

It was, as usual, a perfect California day. The road rose and fell, now down in the shade of redwoods, now up on a ridge with oaks and manzanitas. From the highest point I could see across the small dark-green mountains to the Pacific Ocean. The waters were softened by mist, and on the high, distant horizon was the dark finger of the Monterey peninsula.

I found Carlita's Mexico City–style Mexican restaurant easily enough. It stands right by an old concrete bridge where the state highway crosses the stream called Boulder Creek. I parked and went inside; the place was nearly empty. No sign of Frank Shook. A waiter offered to seat me, but I decided to wait outside. It was about 12:50.

I sat there on a bench in the sun, looking things over. Across the street were some other establishments. The Heron's Healing Beak: Acupuncture and Herbal Medicine. Pot of Gold: A Coffeehouse & More. Four young men were playing slow-paced basketball in front of the San Lorenzo recreational hall. Next to that was the San Lorenzo Fire Department, with an iron bell hanging from a low-mounted log crossbar.

Further up the street was Wilson's Super, a beat-up old supermarket with a neon liquor sign.

A few San Lorenzo citizens passed by. I was feeling keyed-up, and took close notice of each and every one of them.

First came a grunger woman with cowboy boots and round granny shades. She wore a tie-dyed stocking cap, a striped jersey dress, and a sweatshirt tied around her waist. Her white legs looked very bare coming out of the scalloped boot tops.

Next was a big-stomached young man, carrying a baby in a back carrier. He was bearded and ponytailed, perhaps a computer hacker. There's a lot of programmers who telecommute from the low-rent Santa Cruz Mountains.

Some geezers swung a big white American car into the Carlita's parking lot. In the backseat of the car were suits and dresses hanging from a suspended rod. The couple got out and walked across the lot to the restaurant. He: plaid shirt, jeans with stitching, gray helmet-hair. She: white hair, kind face, pink turtleneck, blue sweatpants, sneakers. They looked like they might be itinerant New Age healers.

More time went by.

A county bus pulled up. A burr-cut boy in a red T-shirt eagerly leaned forward to get aboard. Meanwhile, a longhaired boy was unlashing his bicycle from a rack on the front grille of the bus. Six teens got off the bus in a pack, lighting cigarettes and peering through slitty sunglasses.

Here came a compact car driven by an old woman sitting low down in her seat and barely able to see over the steering wheel through her thick glasses. Something about her mild, round chin made me imagine her saying, "Good golly."

The next car was a whipped-to-shit van—all that was visible inside was a big beard, a nose, the brim of a mesh-back hat. A tough mountain hippie. Even tougher was a hugely mustached man in a green vintage car, a fixed-up, tricked-out classic with fender skirts. He was chewing gum a mile a minute. There's a lot of illegal methedrine labs in the mountains; I wondered if he was from one of them.

Now, all of a sudden a very striking trio of people appeared: a beautiful dark-skinned woman and a handsome man—both holding hands with a short, androgynous figure whose sex I really couldn't decide upon.

The three of them were dressed in odd, shiny clothes. It crossed my mind that they might be saucer cultists. Friends of Frank's? I must have been staring at them too hard, for they paused and stared back at me, giving me such a cool and thorough looking-over that it felt like an insult. I smiled and nodded awkwardly; they ambled away.

Across the street the basketball kept bounce-bounce-bouncing, with the four players slowly gangling around. No hurry. But where was Frank Shook?

A bearded man with dark curly hair walked past me and started to do something to the public phone in front of Carlita's. He wore black cutoff shorts, a purple T-shirt, a black baseball cap, and he was carrying a white envelope in his mouth. There were lots of things hanging from his belt: tools and a portable phone? For a minute I thought he might be a particularly brazen phone phreak stealing long-distance time. But then he walked down the street and got into a van—aha!—Pacific Bell. In the Santa Cruz Mountains, even the people with straight jobs look weird.

It was 1:10 now. I checked inside Carlita's one more time. Frank still wasn't there. But as soon as I came back out, I bumped into him. He looked even thinner than the last time I'd seen him, his skin more dry and leathery. The lenses of his glasses were smudged; the eyes behind them bright and odd. His hair had gone a bit gray and could have used a washing.

"Hi, Rudy, am I late?" He smiled and shook my hand. "I wasn't sure if you'd be here." We went into the restaurant, got an isolated table in the corner, and ordered some food. Salad, chicken soup and iced tea for me; an enchilada platter and a Dos Equis beer for him.

PARATIME

"So, where to begin?" said Frank.

"Let's start with the aliens," I suggested, setting out my fountain pen and my paper. I like to use a good pen; it's a writer's little extravagance. The paper was an ordinary pad of blank unlined paper that I'd gotten from school. "What is it like when you see them?"

He smiled nervously. "I'm still getting used to the idea of telling you about this." His Adam's apple bobbed.

"You've never told anyone?"

"Oh, I've told lots of people, but never a scientist like you. A skeptic. Someone who's going to try and analyze the hell out of it. If I tell the average person in San Lorenzo that I've seen aliens, they're not all that excited. Usually they say they've seen aliens, too. But when I listen to their stories, I can tell that they only imagined it. I only know one other person who sees aliens like me. Peggy Sung. She's this very grasping, materialistic woman who lives just down the road in Benton. But for God's sake, let's not start in on *her*."

"So tell me how it is when you see the aliens."

Frank took a deep breath, looked around the room, exhaled. "Time stops. And they appear. I have my adventures with them, and then they put me back where I started and time starts up again."

"Time stops? Does anyone else notice?"

"My wife Mary can tell when it happens, but *you* probably wouldn't be able to. There's a tiny little glitch in the continuity, but you have to know how to look for it. They could come for me right now. I'd just be sitting here with you aaand—" He moved his right hand slowly through the air "—time would stop and I'd go away and then I'd be back here finishing up my sentence. At least I *might* finish my sentence. If I happened to remember what I'd been talking about. Which is not all that likely. Some of my adventures are real doozies."

"I don't get what you mean. If time stops, then how can anything happen?"

"Time doesn't stop for *me*. The aliens can make my time axis run perpendicular to regular time. We get into what I call *paratime*. Here, let me draw you a picture." He took my pen and began to draw on my pad. "Think of all space as a point, and think of human time as the line going up the page. Now suppose that at each instant of our time you can find a different time direction that goes off at a right angle. A perpendicular paratime axis for each moment of human history."

Just then the food arrived. Frank kept right on drawing, adding more and more detail, moving the pen with an easy, practiced hand. While I watched him, I began to eat. My chicken soup was thick with meat and vegetables; the salad was fresh and well-dressed. I was enjoying myself.

"What are all those little drawings?" I asked Frank.

"Those are different ways that people have become aware of para-

FIGURE 1: A History of Paratime

time," he answered. "The closest thing our planet has to cavemen today is the Australian aborigines. And the aborigines have this thing they call *dreamtime*. A timeless time that's outside of history. I figure that's paratime. Same thing for the Egyptian pyramids. A pyramid is a kind of time machine, isn't it? And in the Bible where Jesus tells the good thief, 'Today you will be with me in Paradise,' that's His way of talking about paratime. And that medieval mystic Meister Eckhart, he was totally into paratime. All of Man's time is one Now. It's everywhere, if you know how to look."

"Were you already thinking about paratime before you met the aliens?"

"Oh, yeah! I've always been able to step out into paratime. A little bit, anyway. I used to be so dumb—" Frank chuckled and shook his head. "I used to think that it was just me zoning out. But really I was having flashes of perpendicular time. Of course, it wasn't till the aliens

FIGURE 2: 3D Paratime Prevents Worldline Crossings

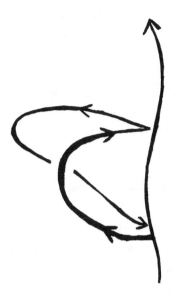

started coming for me that I really got anywhere with it.'' He turned his attention back to his drawing. ''The aliens first noticed Earth after we set off the first atom bomb in 1945. They saw the radiation pulse, and they started coming here, and now they're all up and down Earth's timeline.''

''I don't want to sound like I'm hassling you,'' I said after studying the picture for another minute. ''But what you've drawn here looks like *two-dimensional* time. One regular timeline plus a perpendicular time direction you call *paratime*. How do we get to that *three-dimensional* time you were promising me?''

Frank nodded, took a few bites of his enchilada, and started a new picture. ''Yes. The thing is, Rudy, I simplified my picture so that you could understand it. Of course, there isn't just one direction of perpendicular time, there's lots of them. Has to be. Because when a saucer goes forwards and backwards through paratime, there has to be at least three dimensions of time to keep the saucer from running into its past self.'' He finished the sketch, tore the page off my pad, and passed it to me.

''I see,'' I said. ''And you're saying the aliens can do time travel?''

"Oh, yeah, that's one of the main things we do when they abduct me. We go and look at the future, even though we always stay right here in California. It's a funny thing how we never go anywhere else. The thing is, traveling across space is as big a hassle for the aliens as it is for us. An inch for us is an inch for them, a trillion miles is a trillion miles. And even the aliens can't go any faster than light. But once they actually get somewhere—like the Bay area—they're free to explore the place's whole history. They can visit the past and the future. Going forward and backwards in time is easier for them then flying to a new place."

"So not only can they stop time, they can jump into the past or the future," I said. "Can you tell me any more about how that works?"

Frank began another picture, talking all the while. He was getting really excited. "The way that the aliens stop time is that they hover at one moment of our time by circling around that instant in higher-dimensional time, see. Like a corkscrew or a Slinky." He handed me a drawing and started on the next.

"And when the aliens want to look at our future, they get there by using a shortcut: they take a straight line through paratime that skips over our zigzags. I didn't mention the zigzags yet, did I? Earth's timeline is as shaky as a hound dog sniffing out a rabbit track. Because of quantum mechanics. Earth's time is so crinkly that a thousand years of it is only a few minutes across in paratime; it's like the way you can stuff a quarter-mile of kite string into your pants pocket." The new drawing was already done.

"I like this, Frank." A big smile had crept onto my face. "Though, of course, I don't really believe you. Are you a scientist?"

"Not really. I took physics in high school. I didn't go to college. I watch *Cosmos* and the *National Geographic* specials on TV. And I've read some books. Bertrand Russell's *The ABC of Relativity* and Lancelot Hogben's *Mathematics For the Million*—ever hear of them? And I read a book called *Quantum Reality*—I forget the author—it has two globs like a figure eight on the cover. And of course, there's your book, *The Fourth Dimension*. It's a gas to be talking science with you, Rudy. Though, to be honest, the closest I come to any kind of in-depth technical knowledge is in the area of VCR players and video cameras. Which is how I met the aliens."

FIGURE 3: Stopping Time with Paratime

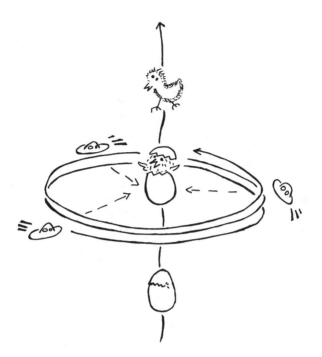

"How did video lead to your meeting the aliens?"

"I'd rather show you that when we go back to the cabin. If we go."
Frank set down my pen and ate and drank in silence for a few minutes.
Finally he looked up and asked, "What do you plan to do with the stuff
I tell you, Rudy?"

This was the question I'd been expecting. I was, in fact, looking for
a new nonfiction writing project, and it had crossed my mind that I might
be able to make a book out of Frank Shook's experiences. I was a little
worried about what might happen to my reputation if I became tarred
with the ufology brush, yet the very perverseness of writing a UFO book
attracted me. And, of course, there was always the chance that it might
make money.

"Well, you haven't really told me very much yet," I said. "But I
imagine you're wondering if I might write about what you say. Would
you mind if I did?"

FIGURE 4: A Paratime Shortcut

"I'd want some of the money. I don't want you taking my ideas and giving me nothing." His face was suddenly grim.

Would this be worth it? What if Frank Shook started claiming that everything I ever wrote from now on was taken from him? "An idea is one thing, but a book is something else," I said carefully. "Everyone has ideas, but almost nobody can write a book. We might be able to make some kind of deal, but if you think you won't feel comfortable about it, then it's probably better if I leave you alone and let you write your own book."

"Listen to the hotshot. Herr Doktor Professor Rucker. All right, how about two percent?"

I'd been bracing myself for a much higher request, and this caught me off-balance. But not totally. "Two—two percent of my income on the book? Domestic or worldwide?"

"Two percent of domestic is fine. But you have to give me all the money you get from—from Finland."

"Two percent domestic plus all of Finland." I said. I looked hard at Frank Shook, trying to figure out if he was putting me on.

"Is it a deal?" he pressed.

"I can visualize it," I said finally. "Yeah. If I sell the book, I could easily give you two percent of my advance."

"And all the money you get from Finland goes straight to me."

"Fine! But later you can't start trying to change the terms and asking for more. And remember that if I don't write a Frank Shook book, or if I can't sell it, then you don't get anything, okay? And no trying to collect from me for all the other books I ever write in my lifetime. No saying that my science fiction ideas come from you. Like, for instance, two-dimensional time happens to be something I've already thought about before, although in a different way, and when you see it in my next SF book, I don't want you getting all greedy and paranoid and trying to bug me about it."

"It's not really the money I'm after, Rudy. I just want you to promise me something so I'll know you're on the level. It's a matter of respect. If you give people information for free, they don't value it. Two percent domestic and all of Finland.[6] It's what my wife and I decided to ask for. And now you said okay, so I guess we have a deal. Are you ready to go to my cabin?"

"Sure." I called the waiter and paid the check.

"I didn't tell you everything about paratime yet," Frank said as we waited for the change. "Your mind can come unstuck from the human timeline and move around in paratime. I think dreams take place in paratime. That's why so many people dream about seeing aliens. Like human time is a big kelp stalk and the aliens are fish floating next to it."

He picked up my fountain pen, tore the last drawing off my pad of paper and started to draw a new picture. I tucked the finished drawings into my briefcase.

"Actually, I think of the saucers as hummingbirds," Frank was saying. "Like a hummingbird sipping nectar from a bottlebrush flower."

A bottlebrush, I should explain, is an introduced Australian plant

---

[6] I never did learn the reasons behind Frank's interest in Finland.

FIGURE 5: Hummingbird Alien

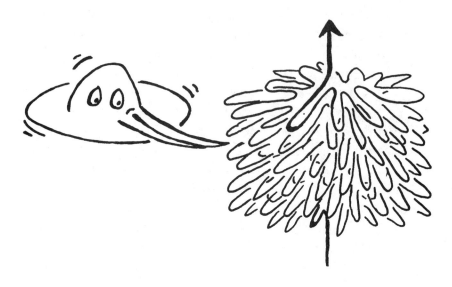

common in California. Its large red flowers are cylindrical spikes with densely-packed radial stamens arranged like the bristles of the kind of brush you'd use to clean baby bottles.

"The alien saucers fly in and out along perpendicular paratime axes to keep probing into the world," continued Frank, still drawing. "But since the world's time is so folded and wadded-up, the aliens can pretty much just turn their beak a little bit and probe into the past or into the future."

"You keep talking about flying saucers," I said. "Are you really telling me that the aliens come in physical metal machines?"

"The saucers aren't machines, they're energy. Of course, energy can look like matter. Einstein says, eh, Rudy? And matter can look like a machine. So anything's possible."

The waiter was there with the change. I left the tip and we went outside. San Lorenzo looked much stranger than it had before.

"You might as well drive," said Frank Shook.

"You didn't drive? How did you get here?"

"Maybe a saucer brought me? Just kidding. Mary dropped me off before she went shopping. Pull out here and take a right."

DRIVING

We drove out of San Lorenzo and took a series of smaller and smaller roads. Frank started messing with my radio, tuning it in on what sounded like static, then turning the hiss down very low.

"One of the big innovations of the twentieth century is speed," he said presently. "Like cars. Driving around in a car changes your relationship to space and time. It's funny how they merge together."

This was an oblique invitation for me to resume my questioning about three-dimensional time, but for the moment I decided to back off and ask something normal. Our luncheon conversation had gotten so unreal so fast that I felt overextended. "What do you do for a living?"

"I work part-time for a toy company that makes a New Age toy called the Lotus Light. It's a little flashlight with a cone attached to the lens; the cone is filled with water and glitter dots. And under the cone there's a turning disk with five colors to change the color of the light. It costs ten to twenty dollars. A high-end product. We're selling them through health-food stores and through sex shops. You're a professor up at San Jose State, right?"

"Yeah. I teach computer science. Object-oriented programming, software engineering, how to write a Windows app, computer graphics, like that. I used to be a pure mathematician, but when I moved out here in 1986, I let the chip into my heart. It feels nice to be teaching something so practical. And of course, I still write science and science fiction."

"Do you think you can sell our book?"

"Maybe. If there *is* a book. It might be worth a try. There's such a big interest in aliens these days. I think it has something to do with the millennium. But, Frank, I have to tell you out front that I'm really doubtful about UFOs. I'm just glad that you're not fingerpaint-the-walls-with-your-own-shit crazy."

"Thanks, Rudy, I appreciate your high opinion of me," he said with a laugh. We were starting to feel pretty comfortable together. "In about a mile and a half you're going to take a left on a little road just after a curve sign. It comes up fast, so be ready. And once we get to my house, I'll try and contact a saucer."

I was getting nervous. "Don't you think your experiences might be dreams, Frank? Or fantasies? Isn't that more likely than there being a whole other order of reality? Now, I'd *like* to believe that there's more to life than the same-old same-old. I'm spiritually dry, and I need to hook into some enlightenment. But I'm skeptical. You're gonna have to help me out with this.''

Frank answered right back. ''Believe me when I say I'm not imagining things, Rudy. I *know* that I've encountered aliens. I've been in the saucers with them, and I've traveled through three-dimensional time. Of course you're skeptical. There's a lot of bad information associated with ufology. My boss who runs the toy company, he was at the Tucson gem show last week, and he said some people were selling things that were supposed to be alien remains from Roswell. I bet they were nothing but dried-up sea skates.''

''Roswell's the place where they were supposed to have had a saucer crash back in the 1940s?''

''Right,'' said Frank. ''But I happen to know for a fact that there's no way a flying saucer ever *could* crash. Any more than a fish could drown. A saucer could no more crash than a breeze could—could shatter. No more than a shadow could blow away. A saucer is like a place where the water bulges up over a rock in a stream. Slow down, here's my road!''

Turning left into this last, small gravel road, I felt a disorienting sense of altered reality—as if I were making a higher-dimensional bend away from normal time.

THREE

Calling the Saucers

FRANK'S HOUSE

The first property on Frank's road was an ugly mess; all of this lot's trees had been lumbered off and the brush had been mowed and chipped. The owners' little house was a shack on stilts above raw red dirt scarred and flattened by tire tracks. Someone's idiotic pioneer fantasy of "clearing the land." There was a bonfire in the middle of the rutted mud. A listless thin-faced man stood by the fire. He stared at us, but he didn't wave.

"He and his family are Okies," said Frank, nonjudgmentally. "The Gandys. As soon as they moved in, they sold off their timber. You can get over a thousand dollars for a big redwood. That's Hank standing there, he's Dick's kid brother. Dick's wife Sharon used to get drunk all the time and come on to me, but then she joined AA. Now it's like slowly she's getting a personality."

We drove on; the forest resumed. The road twisted three times, and then we were at Frank's house, a meager wooden box with tiny windows and an entrance deck of weathered gray boards. There was an old-fashioned aluminum TV aerial on the roof, all spikes and loops.

We parked on the side of the road behind a battered brown Nissan and walked the few yards down to the little deck. The air was a bit hazy with the wood smoke from the Okies' bonfire, but other than that, it felt like being in the heart of a primeval wilderness. Redwoods towered above us, and there were brambles and ferns on the ground. Frank had a few tomato plants growing in buckets covered over with chicken wire to protect them from the deer. There was a rickety picnic table and a wood crate with an upside-down orange cook pot sitting on it.

A dog began barking inside Frank's house, and now the front door flew open and a yellow, snout-faced bulldog came running outside. I stood very still.

"We'll have to see if she likes you," said Frank. "She decides right away."

The dog came to me and sniffed me. She sniffed for a long time—I had the smell of Arf on me—and then she went to Frank for some petting.

"She thinks you're okay," he said, and pushed open the door to go inside. Frank's ugly dog ran back in ahead of us, which I was sorry to see. I'm allergic to dogs in close quarters; I coexisted with my pet Arf by making him always stay outside. As we went in, I noticed that Frank's door had no lock and no knob, only a spring to hold it shut. The bottom of the door had been chewed and clawed away by the dog; it was patched with a piece of plywood.

My senses were tuned to an anxious pitch of alertness. I found myself in what seemed to be a combination bedroom and living room. The air smelled like dog and mold; the lighting was very dim. The main feature in the room was an unmade double bed next to the door, right under the small, curtained windows. Big cushions on the bed indicated that it could also be used as a couch. On the other side of the small room, almost touching the bed, was a spherical wood-burning stove and an armchair. The stove had three welded-on legs and a crooked flue made of black-metal tubing.

"I made the stove out of a buoy that Mary and I found on the beach," Frank said, noticing my interest. "We think the chimney pipe makes it look like a chicken. See the round magnets I stuck on it for eyes? I got those from broken loudspeakers."

A figure appeared in a hall door on our left and Frank greeted her. "Hi, Mary. I brought Rudy Rucker home. Mary, this is Rudy, Rudy this is Mary."

Mary was a slender woman with pale skin and gold-framed glasses.

Her dark hair was pinned up into a casual bun. She wore heavy boots and a green-cotton jumper over a yellow T-shirt. She smiled shyly and waved hello. Her nose was a little crooked.

"Nice to meet you," I said. "Frank's been telling me about aliens and flying saucers."

"Are you going to write the book?" she asked me. "It's all he's been talking about."

"Um, we did discuss it," I said, trying not to sound very committed. "Deciding to write a book takes me awhile. It's not something I rush into." I was worried that Frank wouldn't be able—or willing—to tell me enough that I could use. Then I'd try and drop the project, but he'd keep phoning me to ask about it. Over time the ever-more-frequent calls would go from friendly to impatient to querulous to strident and on into full-blown paranoia about me sneaking around to write this great earth-shaking opus without him so that I could hog all the credit for his brilliant discoveries and—oh man, what was I doing here?

"We're only asking for two percent!" Frank's wife was saying. "That's not so much, is it?" She didn't say this in a challenging way; it was more like she really wanted to know.

"No, no, two percent wouldn't be a problem at all. Frank and I already talked about it, and I think his offer is very generous and reasonable. Um, say, do you have a telephone? I wanted to let my wife know when I got here. She might need for me to come home a little early."

"Frank made us take out the phone," said Mary. "We were getting these like prank calls or wrong numbers from people who'd hang up. And the rest of the calls were from telemarketers. As if we're going to be making all kinds of donations. I don't miss the phone, really." She had a cheerful voice that rose and fell animatedly.

"Peggy Sung was harassing us," said Frank darkly.

"I really don't think Peggy would act like that," said Mary in a reasonable tone. "It was probably telemarketers. I hear they hang up all the time; a lot of the time they're just making calls to find out when people are home. We could have called Pac Bell to put a trap on the line to find out for sure, but you didn't want to."

"Telemarketers don't hang up," insisted Frank. "It was Peggy Sung, I tell you. She's the only one it could have been. The aliens don't make phone calls. And the government doesn't care about them." Frank's attention wandered back to me. "In case you were wondering, Rudy, the government doesn't know jack *shit* about the aliens—and they don't *want* to know. Why would they? Robbing the treasury and getting re-elected is all that matters. And most ufologists don't vote."

"You just think it's Peggy because you have a bad conscience about her," said Mary, stubbornly sticking to her original point.

"I really did want to let my wife know when I got here," I repeated.

"Well, you should have called her from Carlita's," said Frank shortly. "This won't have to take all that long. Let's go into my lab and I'll show you how I get in touch with the aliens. I checked just before I came to meet you, and the Crab Nebula is coming in nice and clear on channel six."

This sounded completely ridiculous. Was Frank Shook perhaps planning to show me a science fiction video and expecting me to take it for a live transmission? He didn't have a phone for me to call Audrey, and my eyes were starting to itch with an allergic reaction from the dog and mustiness. All this just so that I could be harassed about some crackpot book project that was likely to scuttle whatever small scientific credibility my life's work had earned me? "I think I'll go home now," I blurted. "This isn't going to work out. I shouldn't have come."

"Come on, Rudy, lighten up," said Frank. "Would you like a soda?"

"No, thanks." I had images of Mary dosing it with some mountain-hippie brain rot to make sure that I could see the aliens, too.

Frank peered into my face. He looked reassuringly calm and competent. "Rudy. You've come all this way. Let me give you the demo. I know that you're going to like it. I guarantee that in ten minutes you're going to be glad you stayed. Please?" Something clicked in me then and I felt like I was really seeing him for the first time. I decided to trust Frank Shook.

"Oh, all right," I said. After all, I'd come here to experience something new. "Let's do it."

COSMIC SNOW AND VIDEO FEEDBACK

The little hallway off the main room led to a kitchen, a bathroom, and Frank's small "lab"—really just a home office. There was a desk littered with papers, tools, connector cables, and dirty dishes. Singularly for Silicon Valley, there was no computer. But there were three televisions lined up along the wall in front of the desk. A video camera was mounted on an aluminum tripod in front of the television on the left and another camera stood in front of the television on the right. The video cameras were mounted on the tripods in an odd way; the swivel-pan mounts were twisted around so that the cameras were nearly upside down. A third video camera rested on Frank's desk. Some bookshelves filled with dusty paperbacks were attached to the wall behind the desk. The room's only window was covered by a light-tight shade that was pulled down. Frank stepped nimbly forward and turned on the middle TV. It was seemingly not tuned to any channel. Its screen showed dancing black-and-white dots.

"Perfect," said Frank. "Sit down here and watch this with me." He offered me a wooden folding chair, then sat down on a decrepit wheeled chair behind his desk and used a remote controller to turn up the TV's volume. A crackling hiss filled the air: steady, insistent, monotonous, sounding a bit like a waterfall and a bit like frying bacon. The dots continued to dance. The white dots were bigger than the black ones.

"Frank, this is nothing," I said after a minute. "This is video static. When we were little, my brother and I called this a *flea circus*. You could get a flea circus on almost every channel back then. Louisville in the 1950s."

"*Flea circus* isn't the right name," said Frank primly. "It's called *snow*. Did you know that most TV snow is signals that come in from outer space? Just keep looking, and pretty soon there'll be a big wash of extra snow. And I happen to know that it comes from the Crab Nebula."

I was both relieved and disappointed at the method behind Frank's seeming madness. I humored him by continuing to stare at the screen. Every now and then I'd see something like a line that would bend up and form itself into a loop. I made a halfhearted effort to imagine that the lines were forming into outlines of the canonical Gray alien face.

"Let me read something to you while you watch," continued Frank and picked up a book from his desk. "Actual fucking library research. *Is Anyone Out There?* by Frank Drake. He's an astronomer over at U.C. Santa Cruz." And he proceeded to read me the following passage, his voice clear and resonant over the hissing of the TV:

> The Crab Nebula Pulsar remains to this day the only one that alternates its ordinary pulses with giant pulses that are about one thousand times more powerful. In fact, those giant pulses are among the brightest radio signals we have found in the universe. They are so strong you can see them on your television set with an ordinary household antenna. Just turn to a channel that has no program and stare at the screen. Every five minutes you'll get a lot of snow that covers about one third of the screen. That's coming from the pulsar in the Crab Nebula, about six thousand light-years away.[7]

And just about then, as if on schedule, the TV screen did indeed fill with extra snow. "You see?" crowed Frank. "It's true! Not that tuning it in is so easy. Drake doesn't mention that the Crab Nebula signal is at a frequency of around a hundred megahertz. Channel six is eighty-eight megahertz, and channel seven is one-seventy-four. It doesn't have to be that precise, though, so channel six picks up the Crab just fine. You have to choose a time of day when the Crab is high in the sky, and it helps to use an outside antenna."

I looked around the room uncertainly. Overhead were bare rafters and the underside of the low, peaked roof. An antenna cable ran up over the rafters and through a small hole where the wall met the slanted ceiling. I was out in the middle of nowhere and this man was showing me video snow on a television. My confidence was fading again.

---

[7]Frank Drake and Dava Sobel, *Is Anyone Out There? The Search for Extraterrestrial Intelligence* (Delacorte Press, 1992), 91.

"You don't get it yet, do you?" said Frank Shook, catching my expression.

"You're not giving me a lot to go on."

"Well, get ready, Rudy, because now I'm going to tell you. *The aliens are radio waves!*"

"The aliens are the flea circus we're watching on your TV?"

"Yes!" He stared at me provokingly, then threw back his head and chortled. "Professor Rucker is not amused. But wait! I'm not done yet! The aliens are *in* the signals, yes, but they're all encrypted and compressed. My secret is that I know how to get them to unzip!"

He picked up the video camera on his desk and turned it on. The camera had, I noticed, a cable that led to the center TV. Frank jabbed the remote control and now the TV was showing the output of the video camera. He aimed at me, at himself, all around the room. The color was dark and murky.

"You know about chaotic feedback, right?" said Frank. He pointed the camera at the TV, and the feedback generated an endlessly regressing image of a TV screen inside a TV screen inside a TV screen.

"I've done that with my own camera," I said.[8]

---

[8]Here's a quote about video feedback experiments from a wonderful textbook called *Chaos and Fractals*. I've interpolated some additional comments of my own in brackets. The experiment should be set up in an almost dark room. The distance between camera and monitor [and/or the zoom control factor] should be such that the mapping ratio is approximately 1:1. [That is, the image of the TV screen should be just about the same as its actual size. Making the image either slightly smaller or slightly larger than the screen can also produce interesting effects.] Turn up the contrast dial on the monitor all the way and turn down the [monitor] brightness dial considerably. The experiment works better if the monitor or the camera is put upside down. Moreover, the tripod should be equipped with a head that allows the camera to be turned about its long axis, while it faces the monitor. [Even if you don't have a tripod like this, decent results can be gotten using a handheld camera that you hold nearly upside down.] Rotate the camera some 45° out of its vertical position. Connect the camera with the monitor. Now the basic setup is arranged. [If the camera has a manual shutter-speed control, set the shutter speed to a slow value like 1/100 or 1/250.] The camera should have a manual iris [sometimes called the *brightness setting*]

"Yeah," said Frank. "Everybody has. And you probably know what happens when you turn the camera over." As he rotated the camera about its long axis, the screens within screens twisted into a spider-web pattern with a symmetry that shuddered between threefold and fivefold. "I call this a god's-eye," said Frank. "But it's not enough. The next thing is that we need to bring the aliens' signals into the mix. For that I use PIP."

"Pip?"

"Picture in picture." He fiddled with the remote, and now a small preview picture of the staticky channel six appeared in a rectangle in a corner of the TV's screen. Receding small images of the rectangle wound in towards the screen's center. As Frank moved his handheld camera about, the spirals pulsed and warped.

"Now even *this* is something that I'm sure other people have tried," said Frank. "God's-eye plus snow-PIP. Thing is, it's still not enough to get the attention of the aliens. The image is just too simple; it's slightly gnarly, sure, but it's only one level of feedback. To get the aliens' attention, you need to add more into the mix. You need to turn on the other TVs and video cameras and hook everything up into a loop. Turns out three's enough."

"Barnsley fractals," I said suddenly. "You're going to make real-time analog Barnsley fractals. You've seen Barnsley's 'fern' haven't you?"

"No," said Frank shortly. "But if you want to call my patterns

---

which [you should turn down to a low value to keep the screen from being all white, and which] is now gradually opened while the lens is focused on the monitor screen. Depending on the contrast and brightness setting you may want to light a match [or flick a flashlight] in front of the monitor screen in order to ignite the process. [Alternately you can turn the brightness value quickly up and down to get something into the system. Or you can briefly aim the camera away from the TV and at a bright object such as lit doorway on the other side of the room. Or you can use a "picture-in-picture" image from a broadcast channel as a seed.]

—Heinz-Otto Peitgen, Hartmut Jurgens and Dietmar Saupe, *Chaos and Fractals: New Frontiers of Science* (New York: Springer-Verlag, 1992), 19.

FIGURE 6: Three Cameras and Three TVs

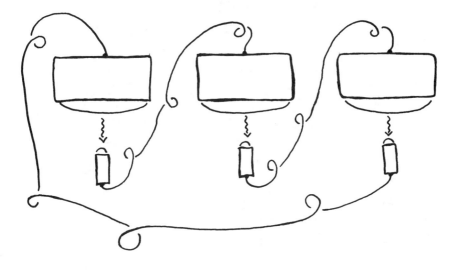

fractals, don't let me stop you. All I know is that the aliens like them. Look, as long as you're all set to analyze this in terms of mathematical logic or something, let me draw a picture of my setup. 'Cause once I get it tuned in, things tend to happen kind of fast—and then I might not feel like talking anymore. Can I have your pen and paper again?''

I handed Frank my pen and the pad of paper. He set down his camera, and started a new drawing. "It's like a daisy chain," he said. "Right now this camera here on my desk is feeding its image into the center TV, but in a minute I'll change the connector so that feeds into the TV on the right. The connector wire from the camera in front of the right TV runs to the TV on the left. And the camera on the left sends its image to the center TV. So there's a loop.'' Frank tilted the paper so I could see it clearly. "And PIP feeds the outer-space signal into the loop,'' concluded Frank.

"Yes, according to how the cameras are positioned you can get all kinds of fractal patterns this way,'' I said.

"Fractals, huh?'' said Frank, still doodling. "They're very organic, very lively. The details look like the whole thing. Have you ever seen a Navajo dream catcher? Rawhide mesh things. Supposed to screen out the bad dreams and preserve the good dreams. This setup of mine is an alien catcher.'' He tore off his new drawing; I put it into my briefcase with the others. I wasn't sure anymore what to expect next.

"What would you like me to ask them?'' Frank said, fixing me with a level gaze.

"You mean ask the aliens? Assuming you talk to them now, that is? Well, if they can show you the future, I'd like . . . oh, I'd like to know what's going to happen to computers. Are we going to succeed in the Great Work? That's a phrase I like to use about what we're doing here in Silicon Valley. The Great Work. Like building a cathedral or a rocket ship. Only it's not obvious to everyone what the goal really is. I think of there as being two main goals: The first goal is for everyone to have full, rich, nearly telepathic communication with everyone else, and the second is the creation of really intelligent machines. Maybe you could ask the aliens to take you into the future to see if the Great Work is going to succeed? Maybe just checking out the first thing would be enough. The future of communication.'' I trailed off, suddenly feeling my request to be absurd. But Frank seemed to take it perfectly seriously.

"All right,'' said Frank, standing up. "Fine.'' He put my fountain pen in his pocket and tucked my pad of paper into his waistband behind his back. "I'll ask them and I'll take notes.''

THE SAUCER DEMO

He walked around his desk and turned on the other two televisions and all three video cameras. He moved the connector from his handheld camera to the right-hand TV and plugged the wire from the left-hand camera into the middle TV.

He fiddled with the two tripod-mounted cameras for a minute, then sat down behind his desk and began wielding his handheld camera. The PIP of outer-space static was still in a corner of the central TV, but the rest of the screen—and of the other two screens—was occupied by in-credibly strange lifelike fractal forms. They looked like plants and in-sects, like spores, like mushrooms, like creepy little frogs.

"I'm gettin' it, Mary!'' he hollered. "It's gonna happen really soon!''

Although Mary didn't answer, I could hear her rattling around in the kitchen. It sounded like she was cooking something.

Frank got more and more excited and then in the space of an instant, with an almost inconceivable abruptness, his expression changed and he collapsed back into his chair, letting the camera fall to the desktop. I would later seem to recall that there was a flickering effect in that instant of change. As if for just the tiniest fraction of a second Frank had momentarily disappeared.

He looked over at me without any immediate recognition.

"Frank?" I said. "Are you okay?"

"Oh, wow, Rudy Rucker, you're still here. Perfect. I've been . . . so far! It feels like it's been days or even weeks. I was all over the future. Finding out about the Great Work."

Footsteps came down the hall; it was Mary carrying a glass of orange soda and a plate of scrambled eggs on a tray. "Here, Frank. Was it a good trip?" She went around to the other side of Frank and leaned over him, taking the items off the tray to set a little place for him at his desk.

"One of the best, Mary."

"Can you tell me about it?" I asked.

"I'm pretty tired," said Frank. "But—I can give you the notes I took. You look at those for a few days and then we'll get back in touch." Moving slowly and wearily he reached behind his back and pulled out my pad of paper. "And here's your pen."

"You better let him rest now," said Mary as Frank handed me the pad and pen.

It was seemingly the same block of unlined paper I'd handed Frank just a few minutes before, but now a bunch of the pages were torn off and resting loose on the top, all filled with Frank's writing and his sketches. The ink looked like the same black ink I use in my pen. Just to check the color, I opened my pen and tried to make a test mark on a corner of the top sheet. Mary watched as I scribbled fruitlessly. Though I'd put a fresh cartridge of ink in my pen that morning, the pen was dry.

Frank was bent forward over his plate of eggs, eating like a starved man, now and then taking deep gulps from his orange soda.

"Please, Rudy," urged Mary. "It's time for you to go."

"No, wait. You can't send me out of here without answering a few questions. Frank! What happened just now?"

He drained his soda with a crackling slurp. "A little more please, Mary," he said weakly. "Bring me the whole bottle."

"He's very tired," Mary said to me admonishingly, and left the room.

Frank gazed at me with a weak, weary smile. "You said you wanted the future of communication? It's all there. Take my notes home and read them. And some other time maybe I can find out about that other 'Great Work' thing you asked about—intelligent computers, wasn't it? But I'll want to research something else first. There's a lot of unanswered questions about the aliens themselves, if I can get them to answer." Frank, yawned, rubbed his face, yawned again. "One thing, Rudy, this stuff is so good, I'm going to have to charge you way more than two percent." Then his head fell down on his chest and he began to snore.

Mary reappeared and ushered me out onto the front deck.

"What do you think?" she asked.

"I'm impressed," I said, riffling through the notes. "These look very interesting. But the incredible thing is that it didn't seem to take Frank any time at all to write these down."

"He wrote them in paratime," said Mary quickly. "In the saucer."

"I wonder," I said.

"You wonder what?" said Mary sharply.

"Well—I wonder if maybe he wrote up the notes before I came here, and then the two of you did a little sleight of hand."

Mary made a wordless sound of outrage, a kind of high-pitched *hmmph*. I backed off a little. "I'm not saying that's what I believe, Mary, but it's the kind of possibility that some people might think of."

"Well, what *would* satisfy you, Mr. Suspicious? If Frank didn't write the notes just now, what happened to all the ink in your pen?"

I shrugged. "Don't worry, if the notes are good, we've probably got the beginnings of a book. I can't wait to read them. How can I get in touch with Frank again?"

"Oh, I'm sure he'll call you in a day or two."

"Okay, then. Thanks, Mary. Nice to meet you. Give Frank my best."

There was an annoying amount of traffic on the way home, but the view from the ridge was beautiful, with the western sky turning yellow and the sun low over the misty Pacific. I tried to bring up an exact recollection of how Frank had looked during that transitional instant at his desk. Had he really dematerialized for an instant? It was hard to be sure.

I had thought Audrey would be anxiously waiting for me, but it turned out she wasn't even home. I'd forgotten to bring my house key, so I had to let myself in with the spare key that we kept in our garbage shed down by the street.

Audrey turned up a few minutes after I let myself in; she'd gone up to San Francisco for the day. She was in a good mood; she'd bought shoes, seen a good show at the museum, and had found a place where she could eat jellyfish for lunch. Like me, Audrey works all winter teaching college classes, and we like to enjoy our summers.

"It was a really dreamy place, Rudy," she told me. "Up on the top floor of this uptight office building in Chinatown where you'd least expect it. There was a big blue-lit tank with the jellyfish in it, and you could pick out the one you wanted. I was completely goofing on the scene. They serve the jellyfish in vinegar and soy sauce." Audrey was a big fan of jellyfish. She liked to paint them and to think about them— and now, even to eat them.

I had fun telling Audrey all about Frank and Mary Shook, and then after supper I spent the evening looking at the notes while Audrey worked on a new painting of jellyfish. She'd done four or five good ones already this year, and the new one looked even better.

FOUR

Notes on the Future of Communication

Reading the first batch of Frank's handwritten notes, I found that each of the sheets was primarily related, if sometimes loosely, to the specific query I'd posed to Frank: What is the future of communication?

This fact alone was quite a surprise. There had only been three or four minutes between when I asked the question and when Frank handed me his notes. Had an invisibly rapid UFO really whisked Frank off to spend a couple of days in perpendicular paratime? This was more than I was yet prepared to believe. I thought it much more likely that Frank had written up his adventures before I even came to his house. And that he'd written up enough diverse topics for his eavesdropping partner Mary to be able to quickly pick out a nice sheaf of pages relating to any question I asked. And that Mary had passed Frank the papers under the tray.

But maybe, just maybe, I was wrong. In any case, the answers were fascinating; they were far richer and more original than anything I'd anticipated. My only disappointment was that I wished the notes had more information about the aliens themselves; presumably that was my own fault for not asking the right question. I resolved that if I did get another opportunity, I'd definitely ask Frank to find out more about the saucerians.

By and large, each page of Frank's notes dealt with a single topic in terms of a set of related scenes that the aliens allegedly showed him. Many of the scenes took place in or near San Jose, the gets-no-respect capital of Silicon Valley, down at the south end of the San Francisco Bay, not far from San Lorenzo. And, with the exception of a few late scenes in South Dakota, the events that weren't in San Jose were in the Bay area.

Later, I would question Frank about this. If the aliens could take him so far into the future, then why didn't they take him to different locations as well? Why not Tokyo, Paris, Dakar, Hanoi, New York, and Rome? Why not Mars and the Moon? Frank had several answers for this. First of all, and most fundamentally, California is the place he was most comfortable with exploring. Secondly, by examining the future of only one region, Frank was better able to perceive the changes that were caused only by the passage of time. And finally, as he'd mentioned at Carlita's, there seemed to be some concrete physical limitations on what the aliens could supposedly do. Although they were capable of taking Frank to other places, doing so would generally cost them more energy than they cared to expend.[9]

Each page of notes included one or more drawings, along with several dozen scrawled words. The brief texts were cryptic and fragmentary; and the meanings of the drawings were not obvious. But Frank was happy to explain them to me.

In preparing his notes for publication, Frank and I decided to print his brief texts verbatim, to use his crude but informative drawings unaltered, and to have me write up accounts based on Frank's rambling but always vivid commentaries on his notes. I should remark that in a few places I've used my scientific background to cast Frank's words into a more technically accurate language than he could be expected to use.

THE LIFEBOX

Lifebox. Big Ad REMEMBER ME! Old man talk, it ask questions. Grandchildren call lifebox Gran'pa. Ask about high school dances—he/it tell about date—ask about

---

[9]A final point relating to saucerian travel: Frank would eventually learn that if the aliens were to leave him off somewhere other than where they picked him up, the trip would have a nasty side-effect.

girl—he tell about her—Sis ask if he fucked—lifebox change subject. Everyone get one, people trade them. Full context. Finally machines can understand humans.

A City Is Like a Lichen

The aliens take Frank into the future, into the middle of the twenty-first century. They're hovering over San Jose, looking down at the city, hanging out near the flight path where the metal airplanes still fly in, the planes looking like saucers themselves from the side; the wings have gotten shorter and wider.

As on his earlier saucer trips, Frank is unable to directly see the aliens. He can't ever seem to turn his eyes directly towards them. It's like they're flickers in the corner of his eye, or as if they're shielded by a blind spot.

They communicate with Frank by projecting voices directly into his mind. The mind merge seems to have a two-way quality to it. As long as he's linked up with the aliens, Frank's brain feels larger and more intelligent than usual.

From the air the city looks like a spreading lichen, an oddly semi-natural growth—Frank muses that people think of a city as an artifact, but at a certain size scale a city is not planned; it obeys the same universal laws of growth as a mold or a fungus.

"Me-shows"

Whenever the aliens want to, they can zoom down to the city and get a closer look at things. As well as looking at real things, Frank and the aliens can pick up signals from TV broadcasts. The aliens can sift through thousands of TV transmissions at once so as to find things that match their current interests.

Frank tells the aliens that he wants to find out more about the future of communication. They begin by telling him that in the future, TV is called *UV*, for *universal viewer*. In the future there are as many different UV signals as there Web pages now. Some of them are just nonstop round-the-clock "me-shows" about individual people.

Frank and the aliens flip through a series of "me-shows." One of them is nothing more than a man driving home from work, watching the long, moving shadow of his car on an evening road, a long California car shadow that crawls over every obstacle like magically stretchable plastic.

Grandpa Ned and the Lifebox

And then the aliens jump to a commercial for something called a lifebox. The slogan is *Remember Me*. The lifebox is a little black-plastic thing the size of a pack of cigarettes and it comes with a lightweight headset with a pinhead microphone, like the kind that office workers use. The ad suggests that you can use your lifebox to create your life story, to make something to leave for your children and grandchildren.

Frank gets the aliens to find an old man who is actually using a lifebox. His name is Ned. They watch Ned from the saucer. Somehow the saucer can use dimensional oddities to get very close to someone but still be invisible to them, even with time running. In addition, the aliens have control over their size-scale and refraction index; they can make the saucer tiny and transparent as a contact lens.

White-haired Ned is pacing in his small backyard—a concrete slab with some beds of roses—he's talking and gesturing, wearing the headset and with the lifebox in his shirt pocket. The sly saucer is able to get close enough to hear the sound of the lifebox: a woman's pleasant voice.

The marketing idea behind the lifebox is that old duffers always want to write down their life story, and with a lifebox they don't have to write, they can get by with just talking. The lifebox software is smart enough to organize the material into a shapely whole. Like an automatic ghostwriter.

The hard thing about creating your life story is that your recollections aren't linear; they're a tangled banyan tree of branches that split and merge. The lifebox uses hypertext links to hook together everything you tell it. Then your eventual audience can interact with your stories, interrupting and asking questions. The lifebox is almost like a simulation of you.

Frank gets the aliens to skip forward to a time after Ned's death. As they do this, Frank is struck by the fact that you can fast-forward past *anyone's* death. We all die, no matter what; it's as fixed and obvious a thing as the fact that each of us has a set maximum height.

Frank gets the aliens to zoom in on two of Ned's grandchildren, who are playing with one of the lifebox copies he left. The aliens are pleased at this zoom, which is not something they would have thought of doing. They really like for Frank to suggest things for them to zoom in on. Otherwise they can't tell what's interesting—they're like humans who

FIGURE 7: The Saucer Watching the Grandchildren

try to have fun watching ants but don't know what to look for. The aliens value Frank for his ability to help them find the significant behaviors. They tell him that he's a much more satisfying kind of saucer passenger than the abductee types who only expect to be humiliatingly masturbated and to have things shoved up their butt.

A lens-shaped little flaw, the flying saucer, appears in the spacetime of a San Jose garage converted into rec room; Frank and the aliens hover there watching Ned's grandchildren: little Billy and big Sis. The kids call the lifebox "Grandpa," but they're mocking it too. They're not putting on the polite faces that kids usually show to grown-ups. Billy asks the Grandpa-lifebox about his first car, and the lifebox starts talking about an electric-powered Honda and then it mentions something about using the car for dates. Sis—little Billy calls her "pig Sis" instead of "big Sis"—asks the lifebox about the first girl Grandpa dated, and Grandpa goes off on that for awhile, and then Sis looks around to make sure Mom's not in earshot. The coast is clear so she asks some naughty questions. "Did you and your dates *do it*? In the car? Did you use a *rubber*?" Shrieks of laughter. "You're a little too young to hear about that," says the Grandpa-lifebox calmly. "Let me tell you some more about the car."

Lifebox Contexts

Frank and the aliens skip a little further into the future, and they find that the lifebox has become a huge industry. People of all ages are using lifeboxes as a way of introducing themselves to each other. Sort of like homepages. They call the lifebox database a *context*, as in, "I'll UV you a link to my *context*." Not that most people really want to spend the time it takes to explicitly access very much of another person's full context. But having the context handy makes conversation much easier. In particular, it's now finally possible for software agents to understand the content of human speech—provided that the software has access to the speakers' contexts.

DRAGONFLY CAMERAS

Like "darning needles." Wings beat in figure eights, never stop.

3 follow a hot starlet to her rendezvous. Ugly boyfriend drives them off. Laser pistol, butterfly nets. Laser bounce off us.

You can see your own news, preview travel. Rent them. Air stability vs. small size. We abduct a gnat camera. Buzzing laughter.

"Why hide?"

Dragonfly News

Frank can't get over the fact that the future TV—the UV—isn't a uniform set of broadcast channels. There are thousands, millions, of UV signals you can tune in with your UV set, which, of course, has powerful computer gear built into it. Frank notices a special new kind of image on a lot of the freelance news shows: Views of stars, criminals, and politicians shot from strange, rapidly moving angles, as if by particularly

FIGURE 8: Dragonfly and Gnat Cameras

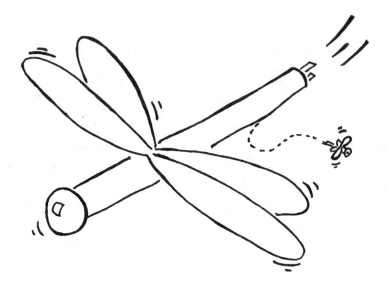

nimble paparazzi. Finally, he hears an announcer mention that one of these scenes is "shot with our dragonfly camera."

The saucer zooms down to find some dragonfly cameras in action. Three of them are following a visiting pop superstar, Milla Maize, who's in San Jose for a concert. Milla is a big sex symbol, and the public is very interested in her doings. The cameras swarm around Milla as she walks from her hotel to her limo.

The cameras remind Frank of the small dragonflies they used to call "darning needles" when he was growing up in Wisconsin. Two or three inches long, with fast-moving wings that never stop beating.

The saucer briefly goes into perpendicular time so as to freeze one of the dragonfly cameras in place so they can get a good look at it. There are four wings, driven by piezoelectric plastic "muscles." The aliens turn their time axis very slightly towards the world's time axis, and now they can see the wings beating in slow-motion figure-eight patterns, sculling the air like two pairs of oars. At the front end of each of the electromechanical darning needles is a tiny camera no bigger than the bead on a glass-headed pin. The lens is coupled to a minute charge-coupled device just like in a video camera.

Now Milla gets into her limo, and the dragonflies dart under the vehicle to affix themselves, leechlike, to the car's undercarriage.

The Dragonfly Paparazzo

While the limo is on the road, Frank and aliens go look at one of the camera operators, who turns out to be an obese cross-dressed man sitting in front of a console in his living room in Milpitas, north of San Jose. His name is Jeremy. With his makeup on, he looks a little like Divine in *Pink Flamingos*. One of Jeremy's friends, a very thin Vietnamese woman, is there talking with him. Jeremy is bragging about his dragonfly; it was quite expensive. Up to a point it's a somewhat autonomous robot—it balances itself and avoids obstacles automatically. But it uses Jeremy's input to decide what to do next. Jeremy watches the camera's view on a computer screen, and directs the camera's motions with spandex sensor-equipped VR gloves. The gloves are gold and glittery.

Milla and the Dragonflies

Frank and the aliens jump to the spacetime location where Milla's limo gets to its destination: a mansion high in the hills above Silicon Valley. Milla gets out and looks around—no dragonflies. But as soon as she disappears into the house, the sly little cameras come buzzing out, circling the house and peeking in the windows.

Later Milla is outside, nude in a hot tub with Carlo, her lover for this evening. Carlo is smart and rich, but he's conspicuously uncharismatic; he's a balding round-shouldered engineer, and not at all the kind of hunk whom Milla likes to be seen with.

Of course, the three nosy darning needles are perched near the hot tub in a bower of jasmine vines, avidly watching. But then Jeremy tries to get his dragonfly a bit closer, and Carlo hears its whirring wings. He snatches up something like a plastic pistol, points it at the sound, and fires an intense pulse of blue laser light, blinding the little cameras. It's a special dragonfly stunner that Carlo's invented for Milla! That's why she's here making love to him!

FIGURE 9: Carlo Zaps the Dragonflies

The next second, Carlo's out of the tub with something like a butterfly net, beating the vines and catching the three dragonflies. Milla pulls the wings off them and Carlo crushes them with a hammer. The aliens love all the excitement. They're happy as myrmecologists catching sight of frantic ants in pitched battle.

For a frightening second, Carlo seems to see Frank's saucer, and shoots a jolt of his light at it. But the light bounces right off the spacetime anomaly, albeit in a strange and disquieting way. Milla starts crying, thinking it's a new and tougher kind of dragonfly. She and Carlo go inside. The saucer follows, and watches them begin to kiss.

Then they dart across town to look at Jeremy; he's crying too. His dragonfly was the chief asset of his fledging *Long Tooth Noser Girl* UV show, and he doesn't have insurance. The aliens take Frank back up to their preferred hover point above the San Jose airport and skip through a few more years, watching for things about dragonflies.

U-Rent-'Em Flying Eyes

Dragonflies stay pretty expensive. A drawback with small dragonflies is that they can't fly very fast. People all over the world have dragonflies up for rental, so that you can log in to some remote site, borrow a local dragonfly, and use it to look around, remotely manipulating it and seeing through its glittering glass eye. Of course, if there's big news in some random spot, then the rates go up and you have to get on a waiting list. But on an ordinary day it's reasonable to rent a dragonfly to take a look around, say, some town you might be interested in visiting. Or perhaps to check out some dangerously sleazy action, like in the alley behind the Will Call Bar in San Francisco's Tenderloin, where Frank and the aliens observe a veritable swarm of tourist-driven dragonflies watching a never-ending parade of sex, drugs, and debauchery.

Gnat Cameras

For ultra-invasive snooping, some dragonflies carry gnat cameras on their underbellies; they can spawn off a few gnats and send them in through a house's ventilation system and, if all goes well, right under a star's sheets. A gnat camera's flight range is so small, though, that it needs a big dragonfly to ferry it around.

Alien Brain Etching

Just for a goof, and with much audible grab-assing, the aliens take one of the gnats aboard for fifteen minutes, and they skip a few days into the future to watch the ensuing UV news reports about the gnat-cam videos of a flying saucer's interior.

The aliens have always prevented Frank from being able to see them directly, so he's very excited to see the UV feeds. The videos show a crazy-looking man—that's our Frank!—with a trio of aliens always standing behind him. No matter how the man in the video moves, the aliens remain out of his view. And what do they look like to the gnat-cam? It's a big letdown. For the purposes of stirring up the human ant-hill, the aliens have formed their bodies into the most obvious mass-media archetype.

In other words, they're disguised as Grays: about the size of children, thin and spindly, with big, bald heads and enormous slanting eyes, with noses, ears, and mouths that are nonexistent or rudimentary. The

Grays look like creatures evolved to do nothing but watch; it's as if they think and see, but do not taste, smell, speak, or listen.

As they watch the agitated newscasters slavering over the gnat-cam Grays, the aliens' laughter sounds like crickets chirping. Frank isn't actually positive that it *is* laughter, that's just what he assumes.

He asks the aliens why they are always so sneaky, and why don't they just come out and, like, land on the White House lawn and meet openly with humanity? Why forever be skulking around?

Their answer is shrill and discursive. It's a super-intense ray of information playing across Frank's head like a dentist's drill—it feels like alien ideas being etched right into the bones of his skull. Frank is sorry he asked, he can't remember what he did ask; oh yeah, he asked why the aliens hide. Their answer: "We don't want people to ask us things."

PIEZOPLASTIC

Plastic wires, plastic batteries, and then it can move. Piezoplastic made of beads. Like a jellyfish. Eats light.

Sewer slug. Toys, LuvSlugs. Slugskates, bottom ripples. Millipede. Big ones for slugmobiles. Snail Man.

Ugly monochrome change to color. Furniture with live paisley. Limpware hackers.

Soft plastic computer chips. Sluggies. The toaster sluggie.

Precious oil. Polyglass.

Piezoplastic Sewer Slugs

While the gnat-cam is in the saucer with him, Frank manages to catch hold of it briefly and to examine its rapidly-beating little wings. As an inveterate tinkerer with things like broken videotape players, Frank is thrilled by the fact that the gnat-cam's wings have no gears, no linkages, and no worm screws. He loves the concept of plastic muscles. It occurs to him that if people can replace gears with soft plastic, maybe they can do the same with all kinds of machines. He decides to use soft machines as his route into the future of communication.

To find out more about the programmable plastic, Frank gets the aliens to skip around in the first half of the twenty-first century until the aliens find a promising year when there's a lot of UV talk about plastic wires, plastic batteries, and a new material called *piezoplastic*. Piezo-plastic flexes in response to electrical impulses and, conversely, sends out weak electric signals if it is flexed.

The plastic-battery technology is incorporated into piezoplastic; the stuff holds a powerful electrical charge, and it can amplify very small signal voltages into heavy-duty contractions. As if that weren't enough, piezoplastic is solar-powered; you charge it up by leaving it out in the sun or by putting it under a high-wattage lamp.

The first commercial application for piezoplastic is very prosaic: sewer slugs. They crawl down a drain pipe and polish the pipe to a glorious inner sheen.

The sewer slugs have a subtle cellular structure; they are sintered together out of piezoplastic beads—that is, the beads are squeezed into a mold and heated to a temperature that's short of their melting point, but sufficiently high to make the beads stick together. Each bead has a preprogrammed set of response curves specifying the way that it inter-converts electricity and motion, and the parallel interaction of this mass of beads leads to emergent behaviors such as rubbing and crawling.

The LuvSlug

A plumber takes a sewer slug home, cleans it up, files off its belt of abrasive teeth, and lets his kids use it for a toy. Frank and the aliens watch as the delighted children christen their new pet Foo-Foo. The eyeless ''Foo-Foo'' flops up and down flights of stairs. It likes to crawl into patches of sunlight and drowse there. If you hold Foo-Foo, the slug writhes around inside your hands. Sort of like a cat, or a toothless pet rat—or a jellyfish.

It's the aliens who put the jellyfish analogy in Frank's mind. They explain to Frank that Foo-Foo is like the sea nettle jellyfish that he's seen on display at the Monterey aquarium. Jellyfish have no nervous systems at all; they're actually colonies of individual polyps. When one of a polyp's neighbor cells contracts, the polyp contracts too—and then it relaxes—and then for a second or two it's too tired to contract again. First it's stimulated and then it's inhibited. The net effect of the inter-acting excitation/inhibition is circular waves that travel out from the center of the jellyfish to its rim like ripples in a pond. This is, the aliens tell Frank, exactly the same kind of parallel computation that sewer slugs use.

Frank and the aliens skip forward. The plumber leaves his family for another woman, and his abandoned wife—an enterprising woman— uses the plumber's credit line to purchase five hundred sewer slugs. She and her male cousin strip off the slugs' teeth and start selling them over the UV as LuvSlugs.

In the UV ad, the LuvSlug is a small, brown oblong with two bumps on it. The bumps undulate slowly, sensually, and there are folds in the piezoplastic that crease and uncrease. The wife uses it on her neck, the cousin on his foot, the wife on her leg, the cousin on his lower back. "Happy to rub you," says the slug's voice-over. The LuvSlug business takes off.

Newk's Oaktown Slugskates

Frank and the aliens zoom in on a black skater-engineer in Oakland who tinkers with a batch of piezoplastic beads to set the stuff's response rate to nineteen times the rate of the beads used in LuvSlugs. He programs his beads by changing the symmetries of the rungs in the long-chain molecules of the plastic. His name is Newk. He explains to a friend that the piezoplastic's molecules are a bit like DNA; each rung of the long molecule can have one of two mirror-symmetric orientations, which means that he can code up one bit per rung. "That's a lot of work for one bit," allows Newk. "But a giant polymer can have as many rungs as you want."

Newk sinters his beads together to make two plastic sausages that he attaches to the bottom of a pair of boots. And then he's skating across the grass beside Oakland's Merritt Lake on his new slugskates. The blades have ripples that get up on top of each other and bulge out into dozens of little fingers that whip along like the legs of a centipede. On downhill runs, the plastic goes into a circular-flow mode, and when the terrain is level or uphill it runs along on little legs. Newk's Oaktown Slugskates become a national fad.

FIGURE 10: Slug-blade

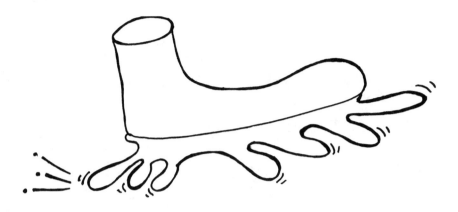

Dustin the Snail Man

The slugs get bigger and faster, and then people start having vehicles like all-weather snowmobiles, with big corrugated slugs on the bottom— the corrugations swing forward and backward in a wavelike motion, trotting across the countryside without damaging the ground.

Frank and the aliens follow along as a paraplegic man named Dustin drives a slugmobile up to the top of Half Dome in Yosemite. Dustin's unwieldy, customized device has metal framework that clanks and scrapes against the rocks. The other hikers are annoyed, though mostly they try to be polite. Dustin, who's a tinkerer and inventor, gets the idea of eliminating every part of the slugmobile except for the piezoplastic. Three months later, he returns to Yosemite wearing a hundred pounds of piezoplastic around his lower body; he calls himself Snail Man. He's found a way to control the plastic with all-but-imperceptible signals from the nerves in his dead limbs. Now he blends into the Sierras as seamlessly as a banana slug.

3-D Paisley

At this point the plastic slugs are still ugly to look at: translucent grayish beige. But then a Santa Cruz woman hacker friend of Dustin's has the idea of throwing some iridescent shimmery stuff in with a batch of piezo-

FIGURE 11: Dustin the Snail Man

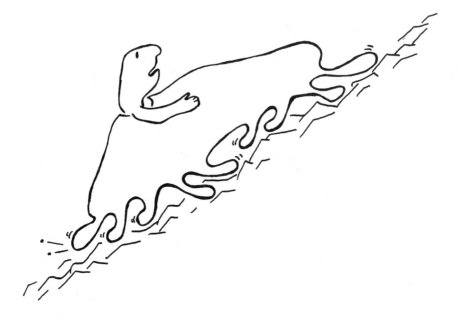

plastic beads and sintering the whole thing together in her oven. Her name is Shirley. Frank and the aliens hover in Shirley's kitchen watching her. Outside it's a sunny day and the ocean is crashing. Shirley has long, carrot-colored hair. The breeze flutters her white kitchen curtains. The shimmery stuff she is adding to piezoplastic beads is rhodopsin-2, a synthetic analog of the visual purple that lines the eye's retina. It's sensitive to electricity. Shirley flops the newly sintered slug out on the table and pokes it. Fuzzy veins of green and mauve bloom within the murky plastic.

The pattern's nice, but not gnarly enough, so Shirley and Dustin set to work tweaking it. Dustin has the idea of mixing a few ounces of generator beads into the sinter mix; the generator beads send out uniform oscillatory pulses. They play with the rates and phases of the pulses until suddenly they get piezoplastic that's filled with a beautiful pattern of living three-dimensional paisley.

Shirley drives Dustin back to his workshop, and Dustin uses his giant

FIGURE 12: Beanbag Chair with Live Paisley

industrial sintering furnace to make up a hundred-pound wad of the new formula. It's gorgeous. Shirley flops down onto it laughing. "I want everyone to have a lifty chair like this! Programmable piezoplastic! Let's farm a world of beanbags, Dustin!"

Chaotic Engineering
Shirley and Dustin dub their new craft "limpware engineering," and begin posting their recipes on the UV as freeware. More and more people start playing with the techniques, and soon the colorful piezoplastic limpware reaches a point where it is doing as much computation as a computer chip. Except it isn't a chip. It's a soft three-dimensional matrix of linked-up cells. And—at least initially—the computations don't perform any "useful" function. They just make pretty colors.

Limpware engineering is a completely chaotic process, as the effects are parallel and emergent, rather than serial and logical. Chaos has its own agenda. The aliens know a lot about chaos, and being linked up with them enhances Frank's understanding of the slippery concept.

One of Frank's alien-catalyzed insights into chaos is that, viewed as a cultural paradigm, chaos means accepting that the half-assed parallel-computed way in which social decisions arise is much more robust and adaptive than any kind of dictatorial guiding could be. The chaos of life means that, willy-nilly, things will tighten around some strange attractor, and you can only hope that the attractor you end up on is where you want to be. And if it isn't what you like—well, you only need to wait a little longer, as the shapes of the basins of attraction are undergoing chaotic evolutions of their own and soon you'll orbit on over to another attractor. Not that this is a very efficient regimen for designing software.

The limpware engineers try lots and lots of different things. One of the slug hackers, a buzz-cut Chinese boy called Jerry, is writing a three-dimensional cellular automata (CA) program to simulate the outcomes of various recipes for piezoplastic—the idea being that its cheaper and faster to simulate, say, a thousand alternate kinds of piezoplastic than it is to physically whomp up a thousand batches.

The 3-D CA displays are intoxicatingly beautiful, though Jerry spends much more time typing in code than he spends watching the images. The aliens stop and watch Jerry for a really long time, much longer than Frank wants to—it feels like two days, maybe—but when Frank complains, the aliens zap him with that skull-etching information ray and he has to shut up.

It's not so much that the aliens are interested in Jerry's 3-D cellular automata, it's that they like observing Jerry in the process of *programming*. They are naturalists of a kind; for them, watching a human hacking is like watching a bird build a nest or watching a spider weave a web.

Texas Machine Language

Even with the speed-up of computer simulations and the ability to automatically try out thousands or even millions of parameter settings, custom-designing limpware piezoplastic is insanely difficult. But then a chemical engineer called Chad breaks through to a high-level programming technique, a way to describe the global behavior you want and to then program that right into an already sintered slug.

Frank and the aliens watch Chad in his Sunnyvale lab. Chad is from Texas. Chad has a contract with a tool company that needs a shock-resistant microprocessor controller for an oil-well drill bit, and Chad's idea is to make the controller out of piezoplastic. He has a complete list of the specifications for the controller, lines of instructions like "when the temperature goes above such-and-such, adjust the angle of the drill-bit teeth to so-and-so." The old solution would be to design a silicon circuit to control electromechanical servomechanisms, but instead, Chad invents programmable piezoplastic.

Each bead of Chad's new-style piezoplastic is made of but a few hundred vast macromolecules, incredibly folded and twisted. Chad has a big diagram of one of them, and the polymer's tangled line reminds Frank of Earth's wadded-up timeline. Each super-molecule includes tiny metallic dipoles, which act as a myriad of antennae, each antenna sensitive to a different frequency of radio waves.

A slug of the new piezoplastic lies there, inert on Chad's lab bench. Input and output wires are attached to the slug here and there with toothy, painful-looking alligator clips. Next to the slug is the parabolic broadcast antenna of a small radio transmitter. Rubbing his hands with callow glee, Chad begins teaching the slug how to act right.

His method is to repeatedly feed a sample stimulus into the slug, meanwhile using his radio transmitter to "tune" the piezoplastic's component molecular types until the slug's physical and electrical outputs are as desired. As soon as the desired response is happening, Chad pounds on the slug really hard with a wooden mallet, which forces the molecules to break any bonds that might have kept them from wanting to stay in their current radio-wave–influenced position. It looks crazy, but it works. Chad hammers his list of specs into the plastic, one after another. The drill-bit controller is a big success.

One of the next applications for piezoplastic is smart door hinges, and from here it's only a short jump to the smart muscles used for the flapping wings of the dragonfly cameras. Soon there are improvements on Chad's "Texas machine language" method of literally beating the program into the slug—and far more complicated applications become possible.

FIGURE 13: Texas Machine Language

Soft Displays

One of the biggest breakthroughs is when a man named Abbott finds a way to use Shirley and Dustin's colored piezoplastic for real-time programmable displays, which are called *Abbott wafers*. All of a sudden, a computer or UV display screen is a cheap, flexible piece of plastic instead of being a expensive, fragile sandwich of glass or a dim bag of liquid crystal. This Holy Grail of ubiquitous computation is finally achieved around 2070. As well as providing cheap, flexible video display, patches of the piezoplastic can vibrate like a speaker membrane.

So now a UV television set is basically just a wad of piezoplastic with a few chips. And then the chips turn into piezoplastic too, and televisions are soft. Some media junkies even use them for pillows. Others wear video clothes.

FIGURE 14: Soft Television Set

Sluggie Processors

The new softly computing limpware-programmed bits of piezoplastic are called *sluggies*. By the late twenty-first century, sluggies have replaced silicon computer chips entirely. Sluggies are sintered from submicron-sized beads, so that the computational density of the limpware becomes as high as silicon's ever was. Just as in the 1990s nobody would dream of using *gears* for the controls of a microwave oven, say, or of a video receiver, in the future nobody dreams of using a silicon chip. All control circuits are smidgens of limpware.

One thing that makes sluggies especially different from silicon chips is that they can move about. And the crawling behavior of sluggies is not a rare or an unimportant activity—no, it's an essential and necessary feature of sluggie self-maintenance. Unless a sluggie gets its daily bit of exercise, its computational circuits fade out and become unusable in a matter of weeks.

This has the benefit that it's very easy to install a sluggie into an appliance. They simply crawl into the device, like a toaster, where they work. But, on the other hand, they also crawl out. Frank and the aliens zoom in on an empty morning kitchen, where the sluggies are all crawling out of their appliances—for the exercise, the air, the light, and just to be with each other. Sluggies communicate via electromagnetic fields, and also by small acoustic chirps.

The sluggies gather together on the kitchen windowsill, lolling there in the morning sun. There are brisk footsteps and the woman of the house walks into the kitchen, dressed and ready for a quick breakfast before hurrying off to work. She wants to turn on the toaster and coffeemaker and the stove—so now the sluggies have to all crawl back. She pokes them with a special sluggie-herding wand to order them back into their machines.

One of the sluggies—the toaster sluggie—is very slow to obey, and the woman gets angry at it. When she comes home from work, she has a limpware upgrade, a new toaster sluggie. She pops it out of its blister pack and sets it down by the toaster. The new sluggie crawls inside, eats the old one—though not without a savage, squealing struggle—and installs itself. When Frank tells me about this, I think of how when you upgrade software on your computer, the new software writes over the old one's directories.

Smart Furniture

A few years later, nearly everything a person owns is at some low level alive, made of piezoplastic that knows what to do. Most of the objects in a person's home can talk a little bit, and for awhile there's a fad for making pieces of furniture with the intelligence of, say, a dog. They get out of the way if you're about to bump them. They adjust their shape to whatever you say. They can change their patterns to match any design that you show them. But smart furniture turns out not to be a good idea.

Frank and the aliens watch as a photographer's family comes home from a week's trip to find that the furniture has been bouncing around the house laughing and bathing its tissues in the studio's klieg

FIGURE 15: Kitchen Sluggies

lights, breaking all the dishes, and running up a huge electrical bill. Yes, the photographer steps into his harshly-lit studio and catches his furniture going wild. A rambunctious over-amped armchair is howling like a coyote, the sofa is galumphing around in pursuit of a long-legged tea table, the sideboard is dancing a tarantella on shards of broken crockery, and six dining chairs are clambering on top of each other to form a pyramid. He loads the rogue furniture into a truck and hauls it off to Goodwill.

In another home, a young woman's disgruntled suitor kicks one of her chairs across the room—and the chair runs back and breaks the guy's leg. A cat sharpens its claws on a couch, and the couch flings the tabby out the window. After a few more incidents like this, the manufacturers go back to making furniture stupid again—though people still like for it to be made of piezoplastic.

Polyglass

As more and more plastic is used for colorful limpware goodies, it becomes less and less acceptable that oil be made into gasoline and burned. Petroleum is so much more valuable if it's turned into plastic. Burning oil for fuel is now considered as wasteful as, say, feeding haute cuisine to a barnyard pig. There are only a few rare internal-combustion car-driving ranges; oddly enough, one of them ends up being Big Sur, the home of so very many auto commercials. But most people have electric cars with slug feet.

Not as much plastic goes into ordinary kinds of things either. Many household items that were formerly made of dumb plastic are now made of special new kinds of glass and ceramics. A man named Junious Gomez capitalizes on the fact that glass is a (very viscous) liquid, and invents a soft, squishy glass called *polyglass*.

Everything in the future is getting softer and less angular. As the new technologies of piezoplastic and polyglass spread across the planet, the centuries-long tyranny of the right angle begins to fade away.

RADIOTELEPATHY

Gruel and water, how many days has it been?

A man on a stage making a light-show with his brain.

Larky and Lucy like two radios vibing off each other. Tweak tweak, fractal filigree of scream.

Lifebox context makes sense. Radiotelepathy and the bigwig. Links to universal viewer. It's a global party.

Perfect computer interface, hack hack. Superanimation.

Record your dreams and play them back. Nelda's dream tape of me and the aliens, disguised as Grays. I think really they're beetles.

Hey, I'm mentioned on future UV and I see our book! This book will make me famous.

Nelda made it stink in here, ugh.

Take me home!

Life in the Saucer

While absorbing all that information about piezoplastic, Frank starts wondering how long he's been out in the saucer with the aliens—not that the question has a clear answer, given that he and the aliens are darting around in higher-dimensional paratime, cutting across the kinks and bends of Earth's timeline in order to see far into humanity's future. The aliens are going all out to research the future of communication. They like having something specific to investigate, and they don't care how long it takes. But Frank's getting homesick and tired.

In terms of his biological clock it feels like he's been off in the saucer for four or five days. He would be starving by now, except that the aliens provide him with an inexhaustible supply of food and water. The food is a kind of porridge in a bowl that never gets empty. And there's an endless mug of incredibly pure water. These items hang in the dark saucer air next to Frank, and when he reaches out for one of them, it jumps into his hand.

The porridge is sweet and exceedingly homogeneous, as if made to a scientific formula for generic human food. Frank thinks of it as "people chow." Frank tries to scrape down to the bottom of the porridge bowl to see where the porridge comes from. At the bottom he can make out little shiny patch, but never for long, because more porridge oozes out a of it whenever it's uncovered. Peering into his cup of water he sees the same kind of shiny glint at the bottom of the glass. Feeling dirty and sweaty, he empties the mug over his head every now and then, and if he looks up into the mug, he can see endless drops of water sweating off the little patch. He can reach his finger up to touch the shiny patch, but it doesn't feel like much of anything, just very slippery, so slippery that in fact he's not sure he's really touching it.

"I want to go home now," he tells the aliens, but they give him the usual kind of skull-etching answer, and plow on forward into the future, flipping through endless UV channels and waiting for Frank to pick up on something.

FIGURE 16: Larky's Brain Show

Larky's Brain Concert

Wearily, Frank focuses on the strange image of a man standing on an empty stage with a tight hood of piezoplastic cupping his head. The hood covers the sides, top and back of the man's head, but not his face. The alien saucer zooms in on this; it's a kind of show being held in a medium-sized San Francisco concert hall. Frank quickly recognizes the hall as none other than the fabled Fillmore, massively retrofitted.

The performer's name is Larky, and the show is called a brain concert. The piezoplastic on Larky's head is seeded with super-sensitive brain-wave sensors, and it's amplifying Larky's brain signals and sending them out to piezoplastic receivers all over the hall. The receivers are giant floppy sheets of piezoplastic lying on the floor and hanging down from the venerable Fillmore balconies. The sheets are ablaze with colors and fuzzy shapes; some of them are recognizable images. In addition, the sheets are writhing, undulating, vibrating, and in fact acting like giant panel speakers, pulsing out hauntingly structured music. Larky is doing all this just by standing on the stage and thinking a certain way into his transmitter-hood. The hood, by the way, is something Larky invented; he's a little like Les Paul or Bo Diddley with their homemade electric guitars.

Telepathy Feedback

Frank and the aliens follow Larky around for a while. The sensors in his hood are off-the-shelf medical-imaging devices, which reach into the brain with tight vortexes of superquantum electromagnetic fields. Nobody's ever thought of using the sensors for art before. "This is the way-perfect tool for me," Larky happily tells his lover Lucy. Up till now he's been a second-rate guitarist who makes his own videos. But his fingers have never been nimble enough, nor his eye quick enough. Finally he has an instrument he can drive with his brain.

Lucy thinks this over and decides that she's in the same boat. "I can't write or paint or play an instrument at all—but I'm a starry thinker, Larky. Can you make me a hood too? We can do duets. And our brains can talk to each other. It'll be so floatin'."

Larky's a little worried about the possible feedback interactions between two piezoplastic transmission hoods, but he and Lucy go ahead and try it.

Curious about what's going to happen, the aliens manage to passively hook into the signals that Larky and Lucy are sending back and forth. They beam the information right into Frank's head, using one of their higher-dimensional mind rays, and that way Frank can interpret the signals for them.

At first it's mellow. Larky and Lucy lie there side-by-side on the floor, smiling up at the ceiling, thinking colors and simple shapes. Blue sky, yellow circle, red triangle. Now Larky puts his hand in front of his face, stares at it, and the image goes over to Lucy. But Lucy isn't able to see the hand yet. She can't assimilate the signal. "You try and send a picture to me," says Larky. He doesn't say the words out loud. Instead, he *imagines* saying them—he subvocalizes them, as it were—and Lucy is able to hear them. Words are easier than pictures. Lucy stares at her bracelet, fixating on it, sending the image out. Larky can't get it at first, but then after a minute's effort, he can. Eureka!

"You have to let your eyes, like, sag out of focus and then turn them inside out, only without physically turning them, you wave?" explains Larky none too clearly, but when Frank tries the technique it works. He too can see through Larky and Lucy's eyes. "It's sort of like the trick you do in order to see your eyes' floaters against the sky," amplifies Larky. "You're looking far away, but you're looking inside your head."

So now Larky and Lucy can see through each other's eyes, but then Larky glances over at Lucy and she looks at him and they get into a feedback loop of mutually regressing awareness that becomes increasingly unpleasant. Frank says it's kind of like the way if you stare at someone and they stare back at you, then you can read what they think of you in their face, and they can read your reaction to that, and you can read their reaction to your reaction, and so on. It gets more and more intense and pretty soon you can't stand it and you look away.

But with a direct brain-wave hookup, the feedback is way stronger. In fact, it reminds Frank of what happens when he points his video cameras at his TVs. Lucy's view of Larky's face forms in Larky's mind, gets overlaid with Larky's view of Lucy and bounced back to Lucy, and then it bounces back to Larky, bounce-bounce-bounce back and forth twisting into ragged squeals.

Lucy and Larky are starting to tremble, right on the point of going into some kind of savage epilepsylike fit—not a pleasant experience for Frank either, as the fucking aliens still have him locked into monitoring the loop!—but Larky does a head trick that makes it stop.

Larky's method for stopping the feedback is like one of the things you can do with the video camera to keep the TV screen from getting all white: you zoom in on a detail. You find a fractal feather and amplify just that. In the same way, Larky shifts his attention to a little, tiny part of his smeared-out mouth, a little nick at the corner, and as soon as that starts to amp up, he shifts over to a piece of Lucy's cheek; just keeps skating and staying ahead of the avalanche. Lucy gets the hang of it too, and now they're darting around their shared visual space.

FIGURE 17: Larky and Lucy

The Mind Modem

Frank and the aliens skip forward through the days, watching Larky and Lucy slowly develop a language for transmitted emotions. Part of the trick is to keep a low affect, to speak softly, as it were. If you scream a feeling, it bounces back at you and starts a feedback loop. You can think a scream, but you have to do it in a calm, low-key way. The way Lucy puts it, "Just go 'I'm all boo-hoo,' instead of actually slobber-sobbing." So pretty soon Larky and Lucy are good at sending the emotions in that gentle chilled-out kind of way.

The aliens occasionally enjoy poking a stick into the human anthill. So, just for a laugh, the aliens patch Frank actively into the Lucy-Larky loop and get Frank to say, "I am a man from 1994, watching you do telepathy from inside my flying saucer." And they send some of Frank's eye images showing the vague, round saucer room he's sitting in.

Lucy thinks it's just a prank that Larky did, but Larky freaks out so much that he's scared to use the equipment again for several days. So Frank and the aliens lay off and let the thing develop further.

Larky actually knows some business guys, and he and Lucy do a demo for them. They go in different rooms and think things back and forth. The business guys try the hoods out, and after some practice they can do it too. It's decided that the word *radiotelepathy* will be the name for this new communication medium. Larky, Lucy, and the business guys form a company called Telepath, Inc., and they get to work developing a commercial product.

For the longest time, there is the persistent problem that if two people use radiotelepathy wrong, they can get into that really nasty feedback loop capable of escalating all the way up to a full-blown grand mal seizure. Not everyone can remember to stay chilled-out and to not stare into the feedback. But finally the Telepath, Inc. technicians develop a chaos-damping algorithm for the hoods' piezoplastic; the damper automatically kicks in to cut off transmission if things get too intense.

The other big hurdle is to make the signals readily comprehensible. Larky and Lucy were able to communicate quite easily because they knew each other really well: They're lovers and best friends. But what happens when you try and link with a relative stranger? None of his or her references and associations make sense.

The trick turns out to be to first exchange copies of your lifebox contexts. As well as using the analog signals of the superquantum brain sensors, you also use standard hyperlinks into the other user's context. The combination of the two channels gives the effect of telepathy.

Frank and the aliens are there at the Fairmont Hotel in San Jose to witness the product roll-out. Telepath, Inc. calls the hoods "mind modems," but nobody likes that name, and the hoods end up being called *bigwigs* on the street. This is not because the hoods are big and funny-looking, but because when you first put one on, the tingling of the electromagnetic vortex fields feels like you have big gnarly dreadlocks or egg-white-stiffened Mohawk spikes shooting out of your head. Bigwigs catch on slowly at first, as they're quite expensive and because people are scared of them.

Wigged In

But pretty soon there are stories about fabulous apps, some real and some apocryphal, and the bigwigs spread far and wide. What are the fabulous apps? Frank and the aliens track some of them down.

The obvious bigwig app is, of course, to make sex better. This is hard because sex is already pretty good without any high-tech help. In some ways people already have all the telepathy they need, what with the multiple channels of sight, sound, scent, touch, facial expressions, body language. If you're in the right mood, you can pretty much read people's minds just by looking at their faces, let alone by making love.

The fact is, to really get an unusual sex thrill out of bigwigs you have to *disable* the damper, which of course everyone soon knows how to do. Some couples become addicted to the dangerous intensity of skirting around the white hole of feedback, of bopping around right on the fractal edges of overamplification, letting their thoughts and emotions bend and whoop. But this isn't for everyone.

The real killer bigwig app comes about when Telepath releases the software protocols and drivers that let users hook their bigwigs up to the global UV "universal viewer" network. Up until then, a bigwig's range was only about a quarter mile. Creating the link is a bit tricky because the bigwig signals are totally analog and UV is digital.

Once you can bigwig together with anyone in the world, the key fun thing becomes looking through other people's eyes. It's like using a dragonfly, but better, because you're looking through actual organic eyes, with a good sampling of the shades of feelings and emotional colorings that come with real experience. And you can talk back and forth with the person watching.

Frank and the aliens observe a guy named Conrad walking on a hill outside Los Perros, California. Conrad's feeling lonely, but he has his 'wig with him and he calls up, just out of the blue, his old friend Ace Weston in Massachusetts. Ace has a bigwig too, and he's walking on the beach, so the two old friends walk together: Conrad on the hill, Ace on the beach, looking through each other's eyes and feeling each other's feelings. It's magical.

The UV-linked bigwig shrinks down from being like a wet-suit hood to being just a floppy, thick patch that you wear on the back of your neck. The price drops way down, and now that they're being used over the UV the name changes: Everyone starts calling the things *uvvies*. *Uvvy* is pronounced soft like *lovey-dovey*. The uvvy completely replaces the telephone and the television set.

The immediacy of uvvy conversations is seductive. People spend more and more time in them; it's an endless interplanetary party that everyone is involved with. It's pleasant and life-enhancing, like you can always plug in with other people like yourself in the country and hang out with them. Being connected via uvvy is called being "wigged in." The diffuse quality of the medium means that you can be wigged into someone without really having any particular message.

Soft Satellites

One of the remarkable things about the uvvy is that there are no uvvy-signal stations, no centralized antenna towers as with 1990s cell phones. Each uvvy acts as a switching point, with signals being handed off from one to the other, just like nerve impulses traveling through neurons. When you buy an uvvy, you don't have to sign up with anything like a phone company or a service provider—at least for local calls. The uvvies handle this themselves.

The catch is that when an uvvy signal needs to cross a big, blank space—like an ocean—it still has to fly up into space to bounce off a satellite. And, for a while, one does need to subscribe to a big satellite service for this. It's evident from the news Frank and the aliens watch that, as of 2080, the satellites are all still owned by what Frank variously calls the Pig, the Man, Big Brother, General Bullmoose, or the Evil Empire—i.e., government and big business. And so long as these ruthless and mercenary forces maintain their control over satellite communications, they continue to broadcast commercial messages and propaganda, to monitor what people do with their uvvies, and to license and control uvvy access.

But in 2085, the final link in truly decentralized global communication is forged. A woman named Ping Wu invents a low-cost communication satellite known as a *skyray*, and by "low-cost" Frank's talking about maybe two thousand dollars. Ping is a high-cheekboned woman with a straight mouth that is often pursed in thought. Frank can't say enough good things about her. Part limpware engineer and part rocket scientist, Ping comes up with a clean and simple design for the skyrays.

They're made entirely of piezoplastic. A skyray is launched as a helium-filled balloon. When the balloon reaches the upper limits of the atmosphere, it splits open and forms itself into a tight cylinder with an ion drive fueled by some newly developed stuff called quantum dots. Quantum dots are a story in themselves, being an incredibly clean and compact source of energy—but no time for that just now.

Once a skyray is at the edge of the atmosphere, its ion drive sends it up into low Earth orbit, where it undergoes a final transformation: It becomes a solar-powered dish antenna, capable of picking up uvvy signals and beaming them either back down to Earth or, for longer hops, to the next skyray in sight. Many people send up a personal skyray of their own, others get clean unmonitored skyray access from some other individual for a small monthly payment.

With the coming of the skyrays, as Frank puts it, "the Pig's pens collapse as fast and flat as the Berlin wall." Wireless electronic communication becomes as unfettered as the conversation among the members of a milling street crowd.

Uvvy Files and Superanimation

The limpware engineers get into using uvvies as the interface to their soft computers. A guy called Omid develops a digital-file format for saving uvvy states. Omid's trick is to use a block of piezoplastic as the recording device; nobody had thought of this before. The soft, richly computing piezoplastic is able to digitize uvvy states down into its individual microbeads. Omid calls his new storage medium the *S-cube* because he loves a woman named Szilvia, though this is a secret he never reveals to the world at large. Frank and the aliens know the secret because they watch Omid telling Szilvia.

Once there is the stable S-cube medium for saving uvvy files, programmers and writers start using uvvies instead of their personal computers. And then the graphic artists get in on the act too.

Frank and the aliens check out a guy called Kotona. He's a *superanimator*. He has this insanely detailed world he's maintaining, doing a kind of waking dream about it every day, building up his database on scores—no, hundreds—of S-cubes.

Kotona's world is nothing less than a re-creation of the world as painted by Bosch and Brueghel. All the set-piece paintings are there: Brueghel's fairs and weddings and feasts and hunting parties, as well as Bosch's allegories and lurid, teeming apocalypses, each painting fully fleshed-out into a three-dimensional scene that—and here lies Kotona's genius—still looks like a great painting from every angle.

Day after day, Kotona enters and reenters his work, ever more smoothly sewing the scenes together into a single, continuous world. Not satisfied to have this world be a static diorama, he breathes life and physics into it, giving the loaves and jugs a mass and inertia, animating the men and beasts to dance and embrace, to whine and to writhe.

Kotona puts his great work—he calls it *Flanders*—up on the Net for any wigged-in user to access. Like a renovating God, he watches his users' movements to find the places where they run into the edges of Flanders, and these are the places where he works to keep adding a bit more, ever working towards his ideal of entering the world of Bosch and Brueghel.

It's a masterpiece, and other artists are inspired to attempt works of similar scope, which are called superanimations. Someone makes an interactive superanimation of Carl Barks's *Duckburg*. Another achieves a superanimation of Picasso's works. And then come artists able to create wholly original superanimations of their own.

FIGURE 18: Kotona's Flanders

Recording Dreams

The next big step in uvvy applications is that people start using them to record their dreams. It's simple. You put on your uvvy, turn on your S-cube recorder, and go to sleep. You can set the uvvy to kick on whenever you get into REM sleep.

Frank and the aliens watch a boring woman named Nelda trying to get her friends to look at the S-cubes of her dreams. "This is wack, Bernice. I'm floating down a river full of, like, lamprey eels and my husband is in this yellow rubber raft with me, he's on all fours crawling around, only he's the size of a baby, and then—*yow*—my teeth start falling out and . . ." Nobody wants to look at Nelda's dreams, except for her therapist, of course, who gets paid to. Frank watches Nelda and the therapist sitting there with uvvies on, floating around together inside of Nelda's most exciting dreams of the week. "That baby-sized thing you say is your husband," says the psychiatrist. "I wonder why you think it is him. To me it looks like perhaps a girl child. Or maybe a lobster."

Just for the anthill-poking fun of it, the aliens give Nelda an intense nighttime abduction experience, making sure it gets taped on her S-cube. The aliens zoom in on a time when Nelda's in bed, and they yank her out into paratime, physically and actually—this is no dream Nelda is having, even though it started while she was asleep.

"Who are you?" Nelda screams at Frank, and points past him towards the aliens, whom Frank himself is unable to directly see. "What are they?" Nelda's awake, though she's so panicked that she isn't acting like a normal wide-awake person. "Go ahead and examine me!" she cries and throws her not-so-appetizing bod down next to Frank. As always, a blind spot seems to cover the aliens, but Frank can make out that one of them is obligingly poking around between Nelda's legs. Nelda loses control of her body functions and then has an orgasm. The aliens set her back in her bed, soiled and wide-awake as she slots back into Earth time.

The S-cube of Nelda's abduction gets a lot of play on uvvies worldwide, and Frank and the aliens watch some of it. The aliens in Nelda's mind-filmed movie look, once again, like short, bald macrocephalic humans in silver *Star Trek* clothes. Like Grays. Frank ascribes this to the images having been filtered through Nelda's ignorant brain, or to a deliberate brain deception on the part of the aliens. In his opinion it's too unlikely that any alien would so closely resemble a human being. He asks the aliens if he's right, and they say yes, the images were a trick. When he asks what the aliens really look like, their answer is a quick image of a luminous scarab beetle—followed by an intense stab of pain to the front of Frank's brain.

One very accurate aspect of Nelda's images is her rendering of Frank's face. It's unmistakably him; Frank's face is on UV shows all over the planet. This makes the second time in this trip that the aliens have caused Frank to be shown to world: First there was the gnat-cam and now Nelda's S-cube! According to Frank, not only do people notice the similarity of the two men in the saucer, but in addition, he claims, they are well aware that these appearances were both predicted by a well-known book that is—as Frank would have it—none other than the book that I'm going to help him write. Frank even gets a glimpse of our book.

Although he would have liked to linger over the images of people discussing him, the heedless aliens rush on to something else, almost as if to torment him.

They tell Frank that now they want to look at some more people's dream S-cubes; they have some theories about human dreams that they want to test. According to the aliens' view of things, dreamers can go directly into paratime, and perhaps even beyond that to some kind of heaven. They want Frank—instead of looking for fascinating news about himself and our book—to help them comb through dozens, scores, hundreds of people's dream S-cubes, searching for images relating to higher-dimensional time.

Frank completely rebels at this. Other people's dreams are boring. And the saucer stinks from Nelda's abduction. And he's exhausted. He screams and hollers until the aliens bring him back home, back to the chair behind his desk.

FIVE

Frank and Peggy

JAHVA HOUSE

As I already mentioned, Frank's notes on the future of communication were no more than drawings with scrawled words when I brought them home on Thursday, June 2, 1994. Actually, the notes weren't quite like they're printed; they were riddled with scores of misspellings. If the truth be told, the first three sentences of the notes went like this: "Lifebox. Big Ad REMEMBR ME! Old man talk, it aks questons." Frank couldn't spell at all.

I read the notes over a couple of times, studied the pictures, showed the stuff to Audrey, and waited for Frank to call. And waited. Just to be doing something, I created a new document in my computer and typed Frank's notes into it; if we went any further this could serve as a starting point for our book. A week went by, and on Wednesday, June 8, he called me around ten in the evening, right after Audrey and I had gone to bed.

"So what do you think of my notes, Rudy?" He didn't bother to say hello, or to introduce himself.

"Is this Frank? The notes are very intriguing. And the drawings are great. I have so many questions. I've been wanting to hear from you. You're going to need to explain what everything means. Should I come back out to your house tomorrow?"

"You don't want to get started right now?"

"I'm in bed, Frank. It's after ten."

Audrey giggled and rolled her eyes as she realized who I was talking
to.

"I thought you cyberpunks were wilder than that! Well, let's get
together in Santa Cruz tomorrow. I have to go down there anyway to
pick up some Lotus Lights. We'll meet at the Jahva House. It's a coffee
shop right off Pacific Avenue, near India Joze restaurant. About eleven
o'clock?"

"Great."

"One reason I'm eager to see you now is that I was out in the saucer
again today. I finally got some good information about the aliens. And
they taught me saucer wisdom. I feel so different. Let me just try and
put it in a really compact form for you. Dig this, Rudy, it's like radio
waves are always around you, and if you happen to have a receiver you
can turn it on, and then bang, there's news and music. The aliens are all
around us too, and they can unzip themselves anytime, but only if they
feel like it. They're like flowers waiting for the right weather to bloom.
And God's all around us too, that's the big thing. You can tune in on
God without any technology at all and God's always ready to appear. In
fact you *are* tuned in to God all of the time, even if you don't realize
it. Everything is alive, everything is God, we're eyes that God grows to
see Himself or Herself or whatever you want to say."

This was pretty heavy stuff to be hearing over a late-night phone
call, but I liked it. As I mentioned earlier, I'm comfortable with the
mystical mode of thought; I'm used to looking through the flakiness of
mystical utterances to see the deep, underlying truths. Since Frank was
talking about God, I asked him a question about religion. "I've been
looking over some UFO writings, and some of them say that, all along,
religious experiences have been in fact visions of the aliens. What do
you think of that idea?"

"Ask him what he meant about piezoplastic and jellyfish," put in
Audrey, who'd been intrigued by the mention of jellyfish in Frank's
notes. I waved her off.

"You have it backwards," said Frank. "The aliens are watered-
down visions of God. Same as you and me. See you tomorrow!"

So on Thursday, June 9, 1994, there I was, punctual as usual, sitting
in the Jahva House waiting for Frank who, as usual, was late. This time
I'd brought my laptop as well as a pad of paper.

The Jahva House coffee shop turned out to be a large old retrofitted garage, a hundred feet square with a high, wooden ceiling with old beams showing. There were fans up beyond the beams, stirring the air around. The walls were coarse red construction tiles, and the floor was cracked slabs of polished concrete with a few threadbare oriental carpets. The tables were wooden, with beat-up brass lamps on them, and the chairs were captain's chairs with wraparound arms.

The coffee bar itself was half an octagon against one wall, and next to it was a little cooler filled with Odwalla fruit juices. I picked out a "Strawberry C Monster" juice, paid for it, and sat down in a chair that was next to a wall outlet where I could plug in my computer. Laminated onto the tabletop was a map of the Monterey Bay. Sitting on the map was a used cup and saucer with traces of hardened cappuccino foam on the cup. I took the cup over to the counter, then sat back down, sipping my juice and looking things over.

The windows were translucent pebbled glass reinforced with chicken wire; they had horizontal pivots in their center so that they could be swung horizontal to let the air in. Out one of the open windows I could see a low, gray building with ventilation ducts on the top and a satellite dish, and above the building were power lines and a pale-blue sky. In front of one of the closed windows hung a stained-glass panel showing a setting sun, a dolphin, and a hummingbird with a flower. Another window showed a green-leafed tree—the window was open, so this was a real tree, not a stained-glass one. The tree's branches were tossing chaotically in the breeze. Inside the coffee shop were half-real trees: potted ficus and dracaena plants, their leaves also moving with the vagaries of the air.

A big speaker in one corner played reggae music. To pass the time while waiting for Frank, I turned on my computer and began typing in descriptions of the scene around me. This is a writerly pastime I learned from reading Jack Kerouac, who called it *sketching*.

A pair of young guys in tank tops were playing chess. One, with curly hair and a purple shirt, sat bent over the board with his back to me. The other was bearded and stoned-out looking, with a turquoise shirt and a black baseball cap worn backwards, his dark hair sticking out to the sides like a wig, his face hardworked and weather-beaten.

Near them sat a dark young man wearing a purple-and-green plaid shirt, baggy torn-off black khaki pants, blue argyle socks, and black sneakers. A big guy, stout. He had a purple stocking pulled over his hair like a snood. His swarthy, bearded pirate face was bent over a picture book from the bookcases that stood along the Jahva House walls. He looked vaguely familiar, but I couldn't place him.

I looked at the other tables. Two young women in granny dresses and no makeup. An old man in a green zip-up shirt reading a paperback. A woman student in polka-dot shorts and a T-shirt studying a stack of Xeroxed notes.

A couple with a little girl caught my eye. He: ponytailed with an earring. She: big shiny eyes, purple tank top, flowered maroon skirt, wise, straight thin-lipped mouth. The little girl had a ponytail and a T-shirt; she walked around stamping her feet in time to the reggae music playing out of the big speaker. It was Desmond Dekker singing "The Israelites." The girl began wrestling her father, tugging at his hand and laughing. She had the same intelligent mouth as her mother.

A man with a complicated wheelchair and a guide dog came in; two kids helped him pull up to a table. A man with sandy-red hair appeared, bobbing his head to the music, his face hidden behind beard and shades. He doctored his coffee at a freestanding condiment bar lined with thermoses of cream and milk, shakers of cinnamon and chocolate, tall glasses holding spoons and straws, and glass jars of brown and refined sugar.

Nobody, other than the little girl, made eye contact with me, nobody hassled me. It was deeply peaceful. But where was Frank Shook? I checked my watch: 11:27.

A young woman with her hair piled on top and wearing a T-shirt with scalloped neck and edges came and talked to the brahs playing chess. She was joined by a man with a toothbrush mustache and a seriously ugly baseball cap, the hat small, perched on the top of his head, and bearing a fragment of a large aloha print of flowers and parrots. Ugly hat got a latte, and scalloped shirt went to the counter and got an iced tea and a chess set; they began playing chess as well.

A man with the suntanned, wind-faired features of a street stoner walked in carrying a conga drum and sat down on a carpet by the big speaker. The pirate-faced man with the purple stocking cap crossed the room, staggering a little, to join the conga man on the carpet.

A sparrow found its way in through one of the huge open windows and flew about.

And then Frank came walking in, wearing dirty white pants and a solid-green sport shirt. He was carrying a green-flannel shirt of a non-matching shade. He spotted me right away, nodded, came over, sat down. He wasn't smiling and his demeanor was abrupt. He seemed to have forgotten all about his God-consciousness of yesterday. I wondered if perhaps he'd been faking a mystical experience just because he knew my interests to lie in that direction. In any case, today Frank's interests lay strictly on the gross material plane.

"First things first, Rudy. Like I said last time, if you're going to be working for me on this book, you're going to have to give me a lot more than two percent of the U.S. royalties. My book is going to be so big. I know for a solid fact that they're still going to be reading it a hundred years from now. You saw that in my first set of notes, didn't you? *This book's success is a foregone conclusion.* I want fifty percent of your worldwide gross. Straight down the middle, half for you and half for me. And you pay your agent out of *your* half."

Although I'd thought I was feeling mellow from sitting there, sketching my surroundings, I got so angry at Frank that I lost control of myself.

"Fuck you," I heard myself saying. "Write your own book, asshole." I powered down my laptop and stood up to leave. "I've got better things to do." My blood was pounding in my temples and my hands were trembling.

Frank looked shocked and upset. "Hold on!" he cried before I could walk away. "Don't be so hostile. Can't we negotiate?"

"Look," I said. "Writing a book means sitting down and thinking really hard and typing—doing it for a long time every day for a lot of days. It's not about *negotiating.* This isn't some Joan Collins book deal on *Dynasty*, Frank. This is real life. If you're going to belittle me and jerk me around, I'm not going to do anything with you at all."

"You think fifty percent would be too much? But all the ideas are mine. You'll just work for me to polish them up."

"You think I'm going to be *working for you?* If you're thinking that way, it means you're going to try and tell me what to do, you're going to breathe down my neck. I can't stand it—I won't do it. My writing is the one and only thing in this fucking vale of tears that I have any control over at all and I'm not about to give it up. Why would I? To write a UFO book with a zero-EQ nut who can't even spell? Forget it!''

The four young chess players had stopped their game to stare at us. The conga stoner said, ''Whoah!'' and the pirate wheezingly giggled. I had been talking pretty loud.

''What does 'zero-EQ' mean?'' said Frank into the silence.

''EQ for *emotional quotient*. Like IQ. *Zero-EQ* meaning no sensitivity to other people's feelings.''

Frank made a placating gesture. ''Oh, sit down, Rudy, and don't fly off the handle. I'm sorry if I insulted you. God, you're touchy. Come on and sit down so I can answer your questions about my notes. The lifebox, the dragonflies, the piezoplastic, the radiotelepathy. Don't you want to hear more?''

Yes, come to think of it, yes I did. Very much so. I took a big, shaky breath to calm myself, and sat back down, heavily sighing. ''First, we're going to have to try and finish what you just started. If we can't agree on some split that you won't want to keep changing, then we can't get anywhere. I'm crazy not to be having my agent do this.''

''But you *do* think it could be an interesting book?'' Frank drew a folded sheaf of papers out of the pocket of his flannel shirt, waggling the sheaf enticingly. ''I've got another batch of notes for you right here. Like I told you, I was off in the saucer again yesterday.''

''What . . . what are the new notes about?''

''Lots of stuff about the aliens. What they're like, how they get here, and what other kinds of life there is in the universe—plus that stuff about God. Don't think it was easy getting this information out of them, Rudy! My head—'' He touched his temples ruefully. ''They were, like, attacking me the way they usually do when I ask them things, but then all of a sudden they decided it would be funny to tell everything precisely *because* we're writing this book. Not that *funny* is exactly the right word.''

He was reeling me in. I had to hear more and he knew it. I sighed again. ''So all right, Frank, let's settle our business and get on with the good stuff. I agree that two-percent domestic isn't enough for your cut, but I also think fifty worldwide is way too much. And you can forget that crap about your cut coming off the top before my agent takes hers. This has to be easy for me or I'm not going to do it. The way it's going

to work is that I give you *X* percent of the checks I actually get. What
we need to do is fix the value of *X*."

"How about a fourth instead of a half?"

"We should put this in writing so you can't change it again."

"I trust you, Rudy," said Frank innocently. "You'll send me
twenty-five percent of every single check you get for our book."

"Twenty. And I don't have to show you what I write before it's
published."

"You don't want my input?"

"I'm going to be getting plenty of input from your notes and our
conversations. But I don't want to hear a bunch of jackass opinions about
the way I put it together."

"You want to be free to make fun of me behind my back, don't
you?"

"Maybe I do! Whatever works. If you want the book to be any good,
you have to let me do it my own way."

Frank suddenly relaxed and smiled. "Fine, fine, fine. Rudy. Do you
realize that I already watched us have this conversation yesterday? From
the saucer. It's floating right up there." He pointed to a spot in the air
between us and the carpet where the pirate and the conga man were
sitting. Catching Frank's gesture, they looked up too, so that now all
four of us were staring at a spot in the air eight feet above the Jahva
House floor.

"What is it, Frank?" called the conga player.

"A little UFO the size of a soft contact lens," said Frank. "How
ya doin', Spun."

"Floatin' oats with Josie Doakes," said Spun cryptically. He got to
his feet and peered up into the air. "I see it. Looks like it's got a little
wobble in the hyperdrive. *Zzzzzow. Zzzzzeee. Zzzzzow. Zzzzzeee.*"

I got up too and peered into the air with Spun. I could almost imagine
that I saw a little ripple in the air . . . but only almost.

"Who you?" said Spun. He had tangled dreadlocks and a long
honey-colored beard.

"I'm Rudy. A friend of Frank's."

"Rudy's going to write my book for me," said Frank. "I had the
aliens bring me here in the saucer from yesterday to watch us make the
deal. So I'd know how much to ask for."

The pirate man stood up too, his purple-stockinged mass of hair bouncing. He stared into the air, his eyes squinting. "I can see the saucer; it's like a little ctenophore jelly with an iridescent sheen on it." He had a slow, rumbling voice. "And I can see Frank inside of it. If you can see today from yesterday, then that makes a paradox. I mean, Frank," he turned his head to look at him, "what if now you do the opposite of what you saw yourself do?"

"Why would I do the opposite?" said Frank. "Rudy's giving me twenty percent worldwide. You've heard of him, haven't you, Guster? Rudy Rucker? He's a famous science and science-fiction writer."

"Um, do you publish under your own name?" asked big Guster.

"Sure." I sat back down with Frank. "You think I'd publish under *your* name or something?"

"What kind of science fiction?" pursued Guster. "Are you into *Star Trek*?"

"I write, like, literary science fiction," I said. "Transreal, freestyle, cyberpunk?"

Guster looked blank, but Spun reacted. "Oh, yeah, yeah; have you ever heard of that writer, um, he's really cool, um—"

"William Gibson." As usual.

"Yeah, yeah, you should write a book like him," counseled Spun, but then a new reggae song came on the loudspeaker, and he and Guster went to sit by the speaker and listen to it, Spun gently tapping on the head of his conga drum.

"How do you know those guys?" I asked Frank.

"I've lived around here for over twenty years," said Frank. "I know everyone. Spun and I used to work for a garden-supply shop together, but of course Spun started growing 'shrooms under the benches in the greenhouse and there was this big bust. You notice how smooth Spun's face is? I always imagine that these stoner types look that way from flying around in intergalactic space. The steady polishing of the interstellar dust. It's much better to fly inside a saucer. And as for Guster—he works for Pac Bell a couple of days a week. He's like a relief worker, he comes in when the regular guys are sick. Or sicker than Guster. He's also a freelance mechanic; I had him work on my Nissan once, which wasn't a good idea. He always has parts left over."

Now I remembered where I'd seen Guster before. He'd been doing something to the phone booth across the street from Carlita's restaurant while I'd been sitting there waiting for Frank last week. Odd coincidence. I brought my attention back to our conversation, back to Frank's claim that there was a tiny flying saucer watching us right now.

"*I* sure don't see any miniature saucer floating over there," I said. "You claim you watched this conversation from yesterday? Can you prove it? Like—can you guess what number I'm thinking of?"

"Twenty percent," said Frank. "But we're done with that, for God's sake. Don't be so materialistic. Did you bring my notes? Spread 'em out and I'll go over them with you. And after that, we can look at the new ones! Oh, wait, one more thing."

"What?"

"I found out the title for our book; it's gonna be called *Saucer Wisdom.*"

"*Saucer Wisdom?*" I said slowly. Suddenly I felt very strange. For *Saucer Wisdom* was the title that I'd whimsically picked for the document into which I'd typed Frank's notes earlier this week. "Where did you get that name?"

"Well, like I mentioned in my first set of notes, I saw our book in the future, but the aliens were rushing so much I didn't notice the title or anything. But yesterday I was off in the saucer again and I got a closer look at the book. I saw a woman who had a copy of it. Over a century from now. And the title is *Saucer Wisdom.*"

"Right." I was still in shock. My mind wasn't coming up with any rational explanations.

"*Saucer Wisdom!*" repeated Frank. "What are you looking so uptight for, Rudy?"

"I—I'm just surprised," I said finally. "I'm surprised because I'd already thought of that title myself."

"Oh, yeah, try and take the credit," said Frank. "But I guess it's kind of a natural title, isn't it? And, that's right; of course you'd think of the same title, wouldn't you? You'd have to. Synchronicity."

For now I'd just accept the madness. "Did you happen to notice who the publisher was? That could really save me some time."

"Um, it wasn't a name I recognized. A real short word. But—I can't remember it."

We spent the whole rest of the day together. We stayed at the Jahva House for another two hours, me typing in all the things Frank told me, and then we went down to the beach and sat there talking some more, with me taking notes by hand. Around five, Frank and I needed a break, so I called Audrey, and then Frank and I went to see *Endless Summer Two* at the Del Mar Theater in Santa Cruz—which was the perfect place to see a surfing movie, as there were lots of surfers in the audience, heartily groaning at the wipeouts and cheering the big rides. After the movie Frank and I talked some more. I didn't get home till 9:30 that evening.

AUDREY

"So how was Frank Shook?" Audrey asked me. She was wearing a happy smile.

"Great," I said. "Crazy. Today I hit pay dirt. I'm bursting with information. Frank told me the most incredible things. It's going to take me a few weeks to write all this up."

"You think he really has a book?"

"It's interesting stuff. People may or may not buy into it as sincere ufology, but I think it could make a nice work of futurism. Somehow our Frank has hooked into a bunch of really wild ideas. *Saucer Wisdom*."

"That's really the title you're going to use?" Audrey had heard me using that name for my file of Frank's notes.

"That's the funny thing. He was already using that title for the book himself. He says he saw people reading it. A century from now."

"Wait a minute. You hadn't told him you've been using that title at home on your computer, but when you saw him today he already knew? He's spying on us, Rudy!"

"He doesn't even know where we live. Our address isn't in the phone book."

"I bet he found out. He nosed around. Or—or he followed you home after your first meeting. Why did you have to get involved with such a psycho!"

"I don't think he's spying on me," I said, willing it to be true. Could

Frank or maybe Mary have followed me home? There'd been a distracting amount of traffic. And I'd been looking at the sunset. I certainly hadn't been watching my rearview mirror. But I didn't want to think that way. "Our coming up with the same title is just a coincidence," I insisted. "*Saucer Wisdom* is kind of an obvious title for this kind of book, don't you think?"

"No. Not to me. But never mind. I don't want to make myself sick worrying about it. Something really good happened today." Her smile returned.

"Tell me!"

"They're going to have a show of my jellyfish paintings in the Los Perros Coffee Roasting Café next month!"

"That's wonderful, Audrey! Congratulations."

"I've been trying to figure out prices to put on the pictures. I already have an idea for what to do with the money. But you might think it's silly."

"What?"

"I want to set up a saltwater aquarium like in that restaurant and keep moon-jellies. They're the easiest. And if that works out maybe I can try comb jellies or sea nettles. Ctenophores, Rudy, the Girdle of Venus! I'll have to get a special kind of filtration system that's really expensive. I found out all about it today, because after I talked to the guy at the café I went straight to the aquarium shop."

"Counting your jellyfish before they hatch."

"I have to entertain myself. With the kids gone and you off spending your days with a saucer nut."

"Don't you love summer vacation?"

PEGGY SUNG

One other thing I should mention about that long Thursday, June 9, 1994, is an afternoon encounter that Frank and I had on the beach.

We were sitting out of the wind in a nook in the rocks to the right of Natural Bridges Beach. Pocked stone sloped down towards the water in front of us; the waves pounded into the rocks and shot up booming sprays of foam. Behind us was a low sandstone escarpment sloping up to a high-tone trailer park. We'd worked our way through all the notes on future communication, and were well into the notes about aliens. I was listening to Frank's dry, pleasant voice and writing a mile a minute on my notepad.

Suddenly he fell silent. I looked up to see a dark figure standing over us, a small-faced Asian woman wearing a white jogging suit and enormous silver-framed sunglasses.

"I think maybe you telling him too much, Frank," she said. "He no pay you yet."

"Peggy Sung! Are you spying on me again?"

"You know it other way around, Frank. He learn my ideas, Mr. Rudy Rucker. Frank promise to be sharing your money with me, not keep it all himself."

"Leave us alone!" said Frank.

"Frank, you should remember who first talk to cosmic spirits," said Peggy.

To my horror, Frank picked up a heavy rock and threw it at the woman in front of us, threw it right at her head with all his might. I slapped my hands over my eyes in horror but—there was no thud, no scream. The rock clattered onto the boulders and bumped into the water. Peggy Sung was nowhere to be seen. I jumped to my feet and peered down off the rocks to make sure she wasn't in the water. No sign of her anywhere. My mind boggled; I felt dizzy.

A high voice sounded from somewhere above us. I peered up, and there, standing at the top of the fifteen-foot slope behind us was the same woman: Peggy Sung. Could she have simply scampered up there in the moment while I'd covered my eyes? "I'll be seeing you again, Rudy Rucker," she called, and then she turned and disappeared behind the skyline. Although this made things even more baffling, seeing the woman whole filled me with relief.

"That wasn't real," said Frank. "And don't ask me about it."

"You try to kill a woman, she vanishes, she reappears on top of a cliff, and I'm not going to ask about it? Come on. Who is this Peggy Sung? I heard you mention her last week. She's your big rival?"

"Peggy Sung is the only other person I know of who contacts the aliens the same way I do. She uses the video-feedback trick, and the aliens take her off in their saucers, just like me. I don't know how they can stand spending time with her."

"I notice you just said '*the* video-feedback trick,' and not '*my* video-feedback trick,' " I said. "Does that mean that Peggy invented it?"

"There's some disagreement on that point," said Frank ruefully. "We came up with it together, but yes, she does say that she invented it. We were both working at the old Western Appliance store down on Pacific Avenue in Santa Cruz about five years ago. I was a stock boy and she was in bookkeeping. We had to work odd hours and sometimes it was just us two. Somehow we got into playing with the TVs and the video cameras. Peggy's very good with machines, though you wouldn't know it to listen to her. Normally she just talks about money."

"And the saucers came and got the two of you at Western Appliance?"

"Yeah. Well, actually, they got Peggy first, and then later she taught me how to get the aliens to come for me. So then Peggy set up shop as a psychic, but I wanted to just keep on exploring. I kind of thought it was a waste for Peggy to be using the aliens to help people see their near future. In fact, I thought she would bore the aliens so much that maybe they wouldn't come at all anymore. I told her not to do it, but she wouldn't listen."

"So there was a fairly strong disagreement."

"Will you stop *coaxing* me? Yes, we had an argument about it, and I told her I'd tell the aliens not to come for her anymore, but she just laughed and said the aliens wouldn't listen to me. And she was right. If I ask them to do something, they ignore me and go on and do whatever the hell they please. It's like they think I'm not worth listening to. Even though I told them it's stupid, they don't even mind helping Peggy with her psychic readings. Thanks to the aliens she's quite good at it, and she makes a lot of money. She calls them *cosmic spirits*. My big fear is that some day Peggy might tell the aliens to stop talking to me. I bet she could do it. And then all my fun would be over. In fact she even got me to promise—oh, never mind."

"And what happened just now?"

"Um—that probably wasn't really Peggy down here. I think it was an illusion. She's done this to me before. She gets in a saucer and has the aliens project images of her. The aliens are good at that; I'm not sure if it's a hologram or a direct brain write. How did she look to you?''

"She was wearing a white jogging suit and big sunglasses. Her hair was up in a bun. She had on pink lipstick and pink nail polish. Her shoes were silver."

"Yeah, that's what I saw too. But you know how I could tell for sure it wasn't real? Her hair wasn't moving in the wind."

"I thought I did see her hair moving. What if you were wrong and she was really here in the flesh? What if you'd killed her with the rock?''

"It would have served her right," said Frank, angrily.

This callous remark marked a turning point for me. From this point on I would always think of Frank as being somewhat unbalanced. No matter how friendly our future conversations might be, I would always keep a little reserve of alertness, a background awareness that this man could turn violent at any time.

"I'm so tired of having her lord it over me," continued Frank, completely unaware of how deeply he'd alienated me. "I could have made the discovery just as well as her. And now that I'm writing this book, she'll be after me nonstop."

"Do you think this Peggy Sung might come after me too?''

"Oh, she probably won't."

"I distinctly heard her say, 'I'll be seeing you again, Rudy Rucker.' ''

"Well, deal with it. Peggy's a burden you'll have to bear, Mr. Eighty Percent. If worst comes to worst you might have to give her some money."

"It sounded like *you're* the one she's asking for money, Frank. I'm not giving up another cent."

"I don't want to talk about it. If I ever promised her something, then I shouldn't have. Let's break and go see a movie. Let's see *Endless Summer Two*. I was a surfer for awhile, you know. Fifteen or twenty years ago, back when I was working as a groundskeeper at U.C. Santa Cruz. It's playing at the Del Mar on Pacific."

So, like I said, we saw the movie and then we went and got supper in the Saturn Café and talked some more, Frank nailing down all the details about his notes on the aliens. We parted on good terms.

On my drive home, I started wondering whether or not I'd really seen Peggy Sung by the ocean. She'd appeared after I'd been sitting there listening to Frank's voice and the sound of the waves for a long time. It wasn't impossible that Frank had hypnotized me, if only for a few minutes. On the whole, this hypothesis seemed more likely than that I'd seen a saucerian brain projection. Yes, hypnosis was a lot easier to believe in. Even easier to believe in was the possibility that there really had been a woman there, and that she had just run up onto the rock slope faster than I'd realized.

I wondered about the nature of Frank and Peggy's relationship. Were they really enemies? Or old friends? Perhaps the beach encounter had been prearranged. Maybe Peggy Sung was Frank's confederate in duping me.

Be that as it may, *Saucer Wisdom* was looking more and more like a really interesting book project. I remained uneasy about Frank's evident mental instability, but I felt that this was something I could work around. It seemed worth the risk, given the fascinating nature of the material he was giving me.

SIX

Notes on Aliens

I was busy for the next two weeks writing up the "Notes on the Future of Communication" chapter and this "Notes on Aliens" chapter. Frank phoned me a couple of times and we had some long conversations to fill in missing pieces. I asked if I could visit him in San Lorenzo again, but he said I should wait until the two chapters were all done. He said if I kept coming to see him I'd just get ahead of myself.

As I worked on the chapters, I went through a bit of a conversion experience. Until I started writing up the notes, I'd thought it quite likely that Frank was perpetrating an out-and-out hoax. But now I no longer suspected that he was deliberately lying. I came to believe that Frank was a little crazy, but fundamentally honest.

It was the very strangeness of Frank's stories that gave them their authenticity. In normal conversation, Frank did not seem like a tremendously intelligent or well-informed person. But here he was coming up with these incredible tales. Either he really was talking with aliens, or he was somehow accessing a deep part of the human mind. It didn't seem possible that he could be just a skillful con man taking me for a ride.

For the nonce, it seemed simplest to assume that the tales were generally true, if only because this made it easier for me to write them up.

COSMIC ROAD TRIP

Tangled disk, AutoCAD city, Las Vegas, the round room. Baggy space, no clear edges. San Jose in front of me, the aliens behind me.

"Let me see you!" Yes! I stare at them, they get closer. Inside-out starfish with spinach and fried-egg. Pukeful. Barbara Stanwyck/Gary Cooper from Ball Of Fire. Keep falling apart, globs, they laugh, I'm scared, they're like boys tormenting a frog. Croak?

I push back: "What do you want from us?" Brain-etching agony. Donald Duck help help. Grit my teeth and push again. "How do you get here and where do you come from?" Black out from the pain. "I'm writing a book! Everyone will read it!" They consult, stop etching, start talking clear channel. I'm a poking stick.

Herman a gnome then a Gray then back to a starfish. They're naturalists, we're dogs. Tourists on a road trip. There's three other saucers visiting me: rainbow roaches, ropy worms and some mystery creatures the starfish won't tell me about.

The Saucer

On Wednesday, June 8, 1994, Frank drinks three cups of coffee to help his concentration, and then turns on his TVs and video cameras to attract the aliens again. On his past trips he's been fairly passive, maybe even a little hypnotized. But this time he wants to be completely focused and aware.

The triple-video feedback plus snow-PIP technique covers the TV screens with fractal, organic-looking shapes. Slowly the shapes seem to thicken, to ruck up and buck out from the curved glass of the television screens. Frank keeps tweaking and tuning the circuits, making the shapes fatter and gnarlier. The patterns pull completely loose from the screens and now there's a cloud of tangled bright shapes floating in front of each of the three TV sets. The dancing knots send tendrils towards each other and link into a single, fabulous form, making a domed translucent disk filled with arabesque filigree. The sound of Mary's radio out in the living room slows down and stops. The rustling of the wind in the redwoods goes away.

Frank sets his handheld video camera down, and the television-screen images collapse down to boring infinite-corridor regresses, but it doesn't matter. The eldritch shape in the room is autonomous now; the

TVs that summoned it are no longer needed. The alien saucer has come again.

The lens-shaped mass of light floats towards Frank, slowly growing. Soon it will engulf him. Frank forces himself to keep his eyes wide open. The bright lines within the growing saucer twitch like someone shrugging, and fall into a pattern of right angles.

Frank took a computer-drafting class in high school; the angular drawings within the alien disk remind him of a wire-frame AutoCAD drawing of a factory—a transparent kind of representation where you just see the edge lines of the walls and ceilings, and the hidden ducts, pipes and wires are all exposed—except that the saucer holds enough dwindling-perspective lines to make up, say, the entire island of Manhattan. And—Frank finds this hard to explain—it's like the lines go in too many directions for three-dimensional space; the lines are all at right angles to each other, but there are more than three perpendiculars; the axes slant off in fourth-, fifth-, seventh-, and tenth-dimensional directions—but before Frank can begin trying to understand it, the saucer is upon him. His skin crawls and prickles as the shape's leading edge moves through and past him, and with the lights surrounding him in every direction it feels for a moment as if he's riding a nighttime convertible down the Las Vegas Strip.

And then—*whoops*—it's like the convertible swerves and Frank tumbles in some inconceivable way out of ordinary spacetime and—*whisk*—he's back in the usual dim kind of saucer-abductee holding cell and—*whoosh*—the saucer flies up through the roof of his cabin, up through the redwoods, and up through the clouds, sloping across the sky to arrive at the saucer's preferred hovering position half a mile above the San Jose airport down at the southernmost tip of the San Francisco Bay, maybe fifteen miles as the crow flies from San Lorenzo and—*whew*—there's Frank looking down at lichenlike "San Ho."

The cell or room in which the saucer carries him has a kind of bench to sit on; it's an amorphous kind of round room made out of force fields or something. There's dim light, and the bright "wiring and plumbing" of the saucer is no longer visible. From the vantage point of this round room, Frank sees right out into whatever section of 3-D reality the saucer is currently pointing at. No matter which way he turns, the 3-D view is always in front of him, as surely as if it were a twin-TV goggle display strapped to his face. What's behind him? The aliens.

Frank's been through all this before. No matter how fast he turns around, the aliens are always still behind him, as if they were monkeys on his back or goblins on his shoulders. What he sees is apparently some kind of direct brain projection or dimensionally warped illusion, some fixed hyper-periscope to whatever spatial cross section of the world's time that the saucer is currently sipping info from. The aliens are always just out of sight, back there in the reticulations of Frank's peripheral vision, part of the prickling on the nape of his neck, unnatural forces perceptible only to his atavistic reptile brain.

"Let me see you!" yells Frank. His voice sounds flat and funny in the *n*-space of the round room. Usually the aliens just read and "write to" his mind; they converse with Frank via radiotelepathy. But today he's got his courage up. Inspired by thoughts of our book project, Frank wants to break through and get some clear-cut answers instead of all the endless, shady equivocal ufological bullshitting around. He wants a clear look at the aliens and he wants definite answers to concrete questions.

The Aliens

In the past, the aliens have reacted to this kind of request with ravaging, punishing mind rays, but for the moment they're quite open. They send Frank a thought like, "All right, you can look at us all you want," and all of a sudden the locked-in view of San Jose comes loose and for once Frank really can turn his head and see something different. He can see what's behind him: three, maybe four, aliens—at first it's hard to be sure how many, because it's not clear where one of them stops and the next one begins.

From all the Hollywood brainwashing, Frank expects them to look like Grays, like big-headed children with stupid little autistic slit mouths and cheesy big schlock-art eyes, but that's not what he sees. And at first he's happy they don't look like Grays, but then he's not. The aliens look like starfish with big, quivery red jelly-eggs in their centers, and with green stuff growing on them in patches like moss or grass. They look flayed, inside out, their flesh is liver-colored and covered with tiny tubules. The force-field barriers between Frank and the aliens drop away,

FIGURE 19: Starfish Alien

and they come more and more clearly into focus. Frank already feels queasy, and then the final veil drops and he smells the aliens. His stomach cramps up, saliva fills his mouth, and he drops down on all fours, vomiting so hard it feels like his stomach is trying to come up out of his mouth.

The vomit disappears into the spatial ambiguities of the round room's floor—if you can call it a floor—but the sharp stomach-acid smell lingers, mixed in with the reek of the aliens. Frank has to rack his brain for words to describe the odor; he comes up with rotten garbage, ammonia, madrona flowers, car-seat plastic, fenugreek, and fresh-baked bread—a combo that adds up to something indescribably loathsome.

Seeing Frank's distress, the aliens wriggle their stubby, starfish arms like Hindu dancers and—just like that—the smell mutates into a pleasant odor of pipe smoke and magnolia blossoms, and their bodies change shape to look like—what the hell?!—the actors in the last movie Frank happens to have watched on television, which was a 1941 Howard Hawks movie called *Ball of Fire* starring Barbara Stanwyck and Gary Cooper—yes, the three or four—or, no it's five—aliens are shaped like Stanwyck and Cooper, three Barbaras and two Garys, all shimmery and black-and-white and moving with that slangy 1940s dynamism, dancing the rumba or something, only—oh-oh!—pieces of them keep coming off and floating around, globs of shimmery humanoid film flesh drifting about the round room, Stanwyck's hips here, Cooper's head there, Stanwyck's hands, Cooper's back, a few of the pieces bump into Frank and they feel warm and real as live flesh, it's so fucking weird; Frank wants to vomit some more, but his stomach is all empty now and he can only heave and retch.

Just to get the hideous body-part rumba over with, Frank yells out a question, the kind of question he knows the aliens don't like: "What do you want from us?" So naturally now they start up the brain-etching thing on him: A beam of pink light shines out of the mass of silvery Cooper/Stanwyck bodies, out of a Stanwyck head that's winking and making a jazzy double-click noise as the beam comes out of her open mouth, the beam tearing into Frank's cortex like a laser into wax, it feels like, oh God, it's dentistry on raw nerves; Frank can't stand it, but it goes on and on.

Somehow Frank has a sudden saving mental flash of Donald Duck in a Carl Barks comic hanging from a tree branch over a waterfall and screaming "Help help!" the way D. Duck always does in a crisis, and this makes Frank laugh somewhere deep inside, and in the momentary forgetting of the pain he finds the strength to holler out more questions: "How do you get here and where do you come from?"

Well, now two more of the brain beams start up and oh Lord, it's too much and Frank blacks out with the pain, then wakes up and they're still tormenting him, leaning over him like ghouls or witches or like cruel children torturing a frog—and with nothing left to lose Frank yells out, "I'm writing a book about you! Everyone will read it! Help me write it, and it'll drive people frantic! My book can be a poking stick for the human anthill!"

This has a good effect on the aliens. They stop working out on Frank's brain and consult with each other. As they confer, their shapes drift into other forms. One looks like a tiger prawn, another like an or-

FIGURE 20: Flesh-Glob Aliens

chid, another like a tornado of sparkling light, another like a quartz crystal, another like a gnome with one single thick leg. Finally they reach a conclusion, and the gnome hops over to Frank. The gnome's mouth reaches entirely around his head, with circular rows of teeth above and below, and when he talks, the top half of his head jounces up and down, exposing a big mushroomlike disk of a tongue, and with several inches of empty air between the appalling top and bottom circles of dentation.

"All right, Frank," says the gnome. "We'll answer some questions. Casting our reality into your . . . your categories will be an interesting

FIGURE 21: Gnome Alien

exercise.'' There's a nasty wet smacking noise each time the top part of the gnome's head plops down onto the big, pillowy tongue.

"Your mouth," says Frank, shakily. "I can't stand it."

"I thought you were a little more original than most people," says the gnome, warping its flesh into the form of a teardrop-faced little Gray. "If it makes you happier, I'll use a body like this." The Gray's short little mouth bends upwards in a sickly-sweet *E.T.* smile. "You can call me Herman." The other aliens form themselves into Grays as well; they stand off at a remove, quiet and luminous.

"Not that shape, either," protests Frank. "It makes me feel so stupid and obvious. Can't you show me your true form? Or did you already do that? Were the starfish shapes the real you?"

"I'm afraid so," says Herman. "Would you like to try again?"

"All right. You just caught me by surprise before. I won't throw up this time."

The alien's flesh flows and shrinks, and Herman take on a shape like—well, actually it isn't really that much like starfish. He does have five main arms—or legs?—but this time the limbs are somewhat different from each other. Each arm is carpeted with smaller appendages, and on the different arms the carpeting is different. One arm seems to be lined with glittering little eyes on swaying stalks, another arm has something very much like a starfish's tube feet, on another is a fuzz of antennae, another is coated with iridescent undulating slime, and the fifth is covered with grasping little hands. Two of the arms—the hands-arm and the slime-arm—branch into pincers at their ends. The bizarre smell returns, but now Frank is braced for it. If he breathes shallowly through his mouth it is bearable.

The most unsettling thing about Herman are the three red sacs at the central meeting of his five arms. The sacs bob and dangle like water balloons, like testicles, like octopus heads, and Frank feels anxious that they might burst.

"Don't worry," says Herman, reading Frank's mind. He thumps on his jelly eggs with his hands-arm. The thumping sets off jolly, resonant vibrations like the pounding of bongo drums. "We're built a lot more solidly than you can imagine. Our race has evolved for ten times as long as yours."

"Oh, you're beautiful," says Frank ingratiatingly. "The more I look, the more I can see how well you're made."

"Indeed," says Herman. "So now I'll tell you some things for your book. It will be amusing to see how your fellow Earthlings react to this information."

Why They're Here

One of the first topics Herman treats is the question of what the aliens want from us. In short, their main interest in Earth and humankind is simply that of tourists looking at interesting animals. They are like whale-watch cruisers, like safarigoers, like naturalists rambling about in the woods. As far as the aliens are concerned, people are fine just as they are. The aliens have no interest whatsoever in teaching, reforming, recruiting, and/or indoctrinating us.

In fact, it's considered very poor taste among the aliens to directly influence the worlds they visit. To make a big public appearance on Earth, for instance, would be a cultural barbarism on a par with kicking holes in a termite nest in a nature preserve.

But it is considered acceptable to now and then pick someone up for a saucer ride, simply so as to get a closer interaction with a native. Generally, the aliens make an effort to only disturb isolated, marginal people whose stories are unlikely to be widely believed.

Owing to the unfortunate contemporary trends in ufology, many of today's abductees expect to be molested and/or masturbated. Now owing to their higher-dimensional abilities, the aliens can see right into a person's body, so they certainly have no practical need to probe into a person to investigate them. If they occasionally do so, it's simply to observe the person's reactions. It's no different than the way someone might rub the stomach of a dog that lies groveling on its back, or the way a completely vulgar person might go ahead and jack off the dog—just to see it writhe and whine.

Although we tend to have the impression that it is one race of aliens persistently visiting Earth, what we are in fact seeing is a small but steady trickle of separate groups of alien tourists. Each alien saucer hangs around Earth for only a limited amount of time—and then goes on to visit other inhabited worlds.

A complicating factor is that the aliens have control over paratime, so that each group's visit to Earth ends up being split up into a few dozen or even several hundred events that are dispersed along the human timeline.

Herman goes on to tell Frank that there are but four distinct saucers that have accounted for the several-score alien encounters that Frank and Peggy Sung have had thus far. There's Herman's saucer of starfish aliens, a saucer of radiant beetles, a saucer filled with creatures who look like coils of manila rope, and a saucer of beings that are—"Well, never mind about them," says Herman. "I don't want to confuse you."

According to Herman, the aliens who visit Earth come from all over the universe. Herman says that the alien method of travel feels essentially instantaneous, so that the voyagers pass by incalculably many-populated planets and stars—as if all at once. Given such a range of choice, there's no sense in trying to stop at every single world—any more than someone on a car trip would get off at every single freeway exit. Each group of alien tourists can pick and choose, finding places that look in some way congenial.

Herman tells Frank that there do exist aliens who are incredibly different from humans; some of them, for instance, are things like sunspots from within stars. But aliens that are so wildly unlike us have no natural interest in our doings—solar-vortex rings can't possibly care about the history of human civilization; instead they would more naturally be attracted to look at what's going on inside our Sun. This means that UFO experiencers like Frank encounter nothing like the full range of possible aliens. The aliens who visit here are beings who in some respect resemble us.

"But this doesn't mean that the aliens who visit Earth are going to be *humanoid*," says Herman. "Only a very primitive being would expect creatures from other planets to physically resemble him or even to have a similar biochemistry. When I say that we resemble you, I mean only that my race comes from a planet, not a star, and that our reproduction method is wetware-based, as is yours. The spectrum of cosmic life is so much wider than the average human realizes."

Frank gets a sudden vision of the aliens' journey as being like a universe-wide road trip where you keep falling into new scenes, no two the same. Incredibly great fun. And finally the aliens are talking openly with him. Frank hardly knows where to begin; he feels like Aladdin in the treasure cave.

"What was that about your travel seeming instantaneous?" he blurts, almost at random. "How does that work?"

"We travel at the speed of light," says Herman. "As electromagnetic waves." Frank is gratified to hear this; it's the answer he was hoping for.

PERSONALITY WAVES

Radio waves, I was right, like faxing themselves. People start to do it in 1000 years. Piezoplastic butterfly wings. Glork money is batteries. 22nd C Turla knows about me and my book. Saucer Wisdom! Personality zettabyte crunches into petabyte S-cube that Turla can mindfax.

4th Millennium, San Jose is gone, grow your house. Lulu's alla creates living matter. What is femtotechnology?

Galactic core. Starfish squeeze out a signal, it snags on our mind-thicket. Seven hundredth stop on a dime. Stripper in the zone.

Fax You!

One of the things that gives Frank his great sense of superiority towards ordinary saucer believers is that most folks imagine the aliens to be solid, meat-type beings who physically fly here in air-filled metal containers— "Like Spam shipped to the Solomon Islands," Frank likes to say. "I mean, why not say that they ride here on horseback? Or sail here in boats? It's ridiculous."

Yes, for a long time already, Frank has thought that the aliens travel as radio waves; he came to this conclusion back at Western Appliance when he and Peggy Sung discovered that triple-video feedback attracts aliens. From this Frank deduced that the aliens are incorporeal ether vibrations from outer space. To his contempt and disgust, Peggy Sung instead drew the conclusion that aliens are cosmic spirits who have infiltrated the TV broadcasting industry! Peggy was also confused about why Frank kept talking about radio waves when it was TV they were looking at. Like many people, Peggy didn't realize that TV signals and radio waves are simply low-spectrum relatives of light waves.

Now Herman the starfish is confirming Frank's thinking. The aliens do indeed travel as electromagnetic radiation that is capable of, whenever necessary, manifesting itself as gross matter. It would be too ridiculously inefficient to send physical bodies across the light-years and parsecs of empty space.

"Certainly none of us would travel great distances in the flesh," says Herman.

Frank compares the process to faxing a picture with a cell phone. Instead of mailing a drawing (read "flesh-and-blood body") in an envelope (read "spaceship"), you code the picture up as bits, send the bits out as a radio wave, and reconstitute the image at some distant location, plucking the necessary info out of the passing electromagnetic vibrations.

"There is one little complication," says Herman. "Although we look like ordinary electromagnetic waves to your limited technology, we're actually coded as higher-dimensional waves. There's a lot more information in our signals than what you see. You might think of it as being like an ice-skater."

"Huh?" says Frank.

"When I travel as a higher-dimensional electromagnetic wave, I'm like a skater resting on a little line of a blade that carves a clean curve across the ice. The blade-line is the ordinary electromagnetic wave that you see."

The fact that the aliens travel as something like radio waves explains why they are tickled and attracted by the gnarly feedback piled upon the snow-PIP signal in Frank's jury-rigged alien catcher.

Soul Broadcast

"Humans will someday begin trying to 'fax' their bodies, just as we do," says Herman. "Look." The saucer lurches, vibrates, and cuts through paratime to intersect the second year of the twenty-second century.

Herman flies the saucer down into San Jose and darts invisibly into a little shop called "Endless You Mindfax."

A woman named Turla sits on a beautifully shaded piezoplastic chair; the thing's patterns are much more subtle and baroque than what Frank was seeing in the twenty-first century's smart furniture. Turla is talking with a customer, a man named Dak. They're far enough into the future that their language sounds pretty strange.

"It's not so waka much to pay," says Turla. "Visualize that it gives you access to the wholo universe."

"How will I know if it very works, Turla?" Dak is old and unhealthy-looking.

"What's to wonder?" says Turla. "We're gonna beam the cosmos a copy of your lifebox context and your DNA code. With lifty tech, some far-off dooks can use the DNA-crypt to grow your bod and then use the context to arrange its mind. You'll be reborn."

"Will I be able to sensey when it happens?"

"You won't know it *here and now*, Dak, but I'm locked that someday . . . somewhere . . . another you *will* know. Everybody very digs that lotsa space-dooks use radio waves for interstellar travel. Aliens are raining down us from the stars, Dak. Deep-space radio signals are crypted life-forms, zipped-up alien derangements."

"How come nobody ever really sees the aliens inside those rays?" asks Dak.

FIGURE 22: Turla and Dak

"The crypt is so lifted that we can't crack it yet. The signals sound like white noise, hissy-hiss. But, Dak, there are plenty wack dooks who say they *have* sensey the aliens."

This is the point where, according to Frank, Turla whips out a yellowed, century-old copy of *Saucer Wisdom* and shows it to Dak, holding it out so clearly that Frank can read the title. To strain my belief the further, Frank claims that Turla reads aloud the Turla-and-Dak passage in *Saucer Wisdom*—presumably this passage right here—and gives a greeting to the invisible Frank and the saucer. When he recounts this tall tale to me, Frank says this is more proof that our book is going to be a big success, and more reason why I should be grateful to him for letting me work on it.

And then Dak is paying Turla; he's giving her three glorks—which are things like coins except they're incredibly small and powerful batteries, each holding several kilowatts of clean energy stored in the form of quantum dots. Energy is an eternally valuable commodity.

Dak puts a very fuzzy piezoplastic patch on his neck, and Turla gets out a green little patch of her own.

"I call this Captain Crunch," says Turla, flicking her piece of green plastic. "It's got mongo throughput." She eyes the uvvy already on Dak's neck. "You've fully put your lifebox context onto it, yes yes Dak?"

"Affirmo," says Dak. "I did it."

"The context makes Captain Crunch work terrif better," says Turla, and puts the green patch of custom piezoplastic on top of Dak's uvvy, followed by something like a little cube of red Jello-O, which she places on top of that. "You going into the S-cube now, Dak. *Whoosh!*"

The uvvy's electrosensitive tendrils feel into Dak's brain, and Turla's custom cruncher compresses the information on the fly, feeding it into the red S-cube.

"We taking a zettabyte of Dak down to a petabyte of crypt," says Turla.

(Note by R.R.: Frank said he distinctly heard Turla use that word: *zettabyte.* I'd never heard of that particular magnitude, but when I looked up the official Scientific Bureau of Standards size prefixes in John H. Conway and Richard K. Guy's *Book of Numbers* [New York: Springer-Verlag, 1996], I found it as shown on the following table.)

NAME	NUMERICAL SYMBOL	PREFIX
THOUSAND	1,000	KILO-
MILLION	1,000,000	MEGA-
BILLION	1,000,000,000	GIGA-
TRILLION	1,000,000,000,000	TERA-
QUADRILLION	1,000,000,000,000,000	PETA-
QUINTILLION	1,000,000,000,000,000,000	EXA-
SEXTILLION	1,000,000,000,000,000,000,000	ZETTA-
SEPTILLION	1,000,000,000,000,000,000,000,000	YOTTA-

And then Turla turns on a small radio-dish antenna and broadcasts the S-cube code for Dak straight up into the sky.

"For your three glorks, I'm sending your crypt out a hundred-fold times. It'll scoot off in a hundred different ups, Dak," says Turla. "I *so* visualize you getting unpacked somewhere somewhen somebody."

Matter Rays

"Turla's wrong about that," Herman tells Frank, "So far as I know, nobody ever decrypts other creatures' personality waves."

"Too much trouble?" asks Frank.

"Of course," says Herman. "Like, do you really think anyone would bother to try and nano-fabricate an alien being's wetware and grow it a body? And decrypt some kind of life-box code? The very fact that you would *need* to decrypt someone's personality wave means that it was made by someone with such low technology they're not going to have anything interesting to tell you. Like, I mean, you don't see humans expending much effort in trying to translate pig squeals into narratives of hog consciousness, do you?"

The starfish twitches and chuckles, continues talking. "You have to make a signal that's self-extracting, and you have to master femtotechnology to extract out into a physical form. You humans won't actually figure that out until what you call your third millennium. Let's take a quick peek."

The saucer heels over to one side and knifes further into the future, though the trip doesn't seem to take any time at all.

San Jose isn't there anymore in the year 3003; apparently the Big Quake has come and gone, quite radically enlarging the San Francisco Bay. People are flying around like giant butterflies, their bodies encased in piezoplastic suits with great gossamer wings. The air is clean, the bay water is pure, pollution is a thing of the past.

There are some nice waterfront cities spreading up the foothills of the Santa Cruz Mountains, and in one of them is a small house covered with flowering vines. A Burmese scientist named Lulu Ma lives here with her husband and assistant Yanno.

On closer inspection, not only is Lulu's house covered with vines, the house itself is made of redwood—only it's living redwood, a great wondrous burl of the stuff that has somehow been coaxed to grow into three or four generously proportioned rooms, complete with cunning knothole doors and windows. It reminds Frank of a house in a picture book he loved as a child.

One of Lulu's rooms is kind of a lab. There's a ledge in the thick redwood wall bearing a variety of small, utterly incomprehensible hand-made devices. One of them has the look of a magic wand, with a business

FIGURE 23: Butterfly People

end consisting of a triad of three mutually perpendicular little rods. Lulu and Yanno are talking, but it's almost impossible for Frank to understand their speech. It's some kind of English, sure, but their intonations and slang are too different. Frank watches them like he's watching a foreign movie, though every now and then a single recognizable word will jump out.

"Blur blur yabba alla now," says Lulu, adjusting the wand thing, which Frank assumes is the *alla* Lulu mentioned.

"Udda yudda zing zang femtotech," says Yanno. Frank's just heard

FIGURE 24: Lulu with an Alla and a Big Tongue

Herman mention *femtotechnology*, but he doesn't know what it is—and nobody's going to tell him yet today.

"Sst sst sst," laughs Lulu and turns on the alla. It's pointed at an empty little stool on the other side of the room. The stool seems to be made of something like glassy black stone with lots of patterns cut into it. The air on the tabletop shimmers and a puddle of pink water appears, now more and more water—or is it blood? It seems to be getting darker in color, and it's running off the stone table onto the redwood floor.

"Gonna wong flaboot!" says Yanno, clearly annoyed. He doesn't like the mess on his floor. He whistles sharply and something like a big pink tongue appears from somewhere, undulating along on its belly like a slug. The tongue laps up the liquid. "Oo bam," says the mollified Yanno.

"Loka moka," says Lulu, waving the alla, and now a growing sausage of scaly green—is it the tail of a snake?—starts coiling and piling up on the stone table. But suddenly the alla runs out of power and the demo comes to a stop. The green snake or lizard tail is impatiently twitching.

"Needa glork," says Yanno.

Herman interrupts. "Let's go back to 1994 before we lose track of your spacetime coordinate, and then I'll show you a mental movie of how we put mind-faxing and matter-transmission together back home."

From the Galactic Core

In another instant there they are above the S.J. airport again, and something a little like an educational science cartoon starts up in Frank's head.

"So now I'll tell you more about how we get here," continues Herman. "I'm enjoying opening up for once. You're a singularly receptive person, Frank. First let me show you where we come from. Our home is a planet near the center of the galaxy." Herman flashes an image into Frank's brain.

Frank sees a big whirlpool of spacetime, a chaotic maelstrom of light, a vast, superheated gas spiral around the bloated black hole at the galactic core.

A few score light-years out from the core is a reddish sun with a bulging blue planet. "This is our home," says Herman. "We call it ♦." The name is a shape and a color, not a sound, but since the conversation is telepathic it doesn't matter. The view zooms in on ♦, and Frank can see legions of the alien starfish, packed as thick as anemones in a tide pool.

♦ is a water world—or is covered with *some* kind of clear liquid, anyway—only not very deep. Frank can see through the stuff to the starfish crawling around on the rocky bottom.

But they are doing more than crawling around, Frank realizes. Looking closer, he can see that the rock that he took for a natural shoal is worked and structured in intricate patterns. There are curving cracks that lead down to lower levels, there are lacework arches, there are even towers that rise above the gently-lapping waves. The starfish are moving about along smoothly plotted paths, propelled by little devices like water jets. Frank's looking down at a submarine city.

"We like to climb into the towers at night to gaze at the galactic center," says Herman. "The great central black hole makes a remarkable lenslike effect against the sky. And the accretion disk is an ever-changing

FIGURE 25: Starfish Aliens on a Tower Beneath a Black Hole Sky

cascade of light. There's some strange energy beings who live in there, you know. They look like corkscrews. We call them the wigglers, but we have very little to do with them. They're not wetware based, so we have almost nothing to talk about. But it was they who gave us the idea for full transmission.''

Frank's looking down at a swaying gossamer spire on ♦. The structure seems to grow as he watches; it's alive. Perched in the top of the tower are five starfish. They touch each other and suddenly there's a flash and crackle as a signal beams up from the tower. This is an higher-dimensional electromagnetic encryption of the creatures' whole selves: body, mind, and soul.

Perhaps a bit too artily, the mental cartoon splices in an image of an Earthly dandelion in full bloom: A sphere of gossamer-white with the breeze lifting off little feathery parachutes, each carrying a seed.

"Yes, yes, throughout the universe, beings send out copies of themselves," says Herman. "It's a universal imperative; replicate or die. Beyond the ordinary wetware-style reproductive copying of self lies the higher, holistic nongenetic duplication. Full transmission. The result? An eidetic copy of the whole self that can leave the home world and join the cosmic net of traveling minds. It's life at a higher level."

The Smell of Souls

"Although the trip is the main thing," continues Herman, "The stops we make along this journey are valuable in and of themselves. Your planet is approximately the seven-hundredth location where we've broken our journey and visited. It's incalculably more primitive than our home, but it's a nicely exfoliated wetware world just the same. And of a very fine chaoticity."

"How do your personality waves know where to dig in and unpack themselves?" asks Frank.

Herman explains that the way they tell if a place is interesting is by looking at the crookedness of the particle worldlines at that venue. Because sentient creatures observe things so sharply, they make the line of time more zigzagged and less smooth. By way of explaining this, Herman asks Frank to imagine a coin being flipped by a dumb, blind machine. If there's no conscious mind to look at the coin, the universe is content to let the line of time near the coin be a bit vague and fuzzed, a kind of half-heads half-tails situation. But if there's a thinking creature looking at the coin toss, then the flow of time has to fork either towards "heads" or towards "tails." The upshot is that the presence of consciousness makes bumpy patterns in the otherwise even flow of time. The patterns are like rumble strips near a tollbooth, or perhaps like trees in a spring-swollen river. Yes, for the aliens, there's a perceptible aura or "smell" about any spacetime location where sentient beings live.

FIGURE 26: Boring Worldlines and Gnarly Worldlines

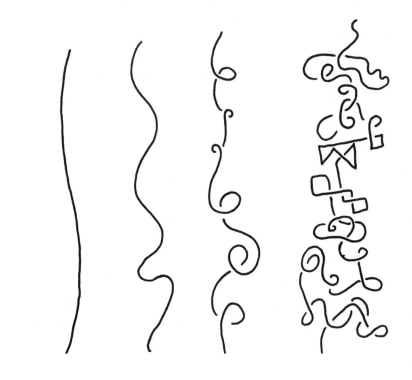

Creating a Saucer

The mental science-cartoon resumes. The light-speed signal of an encrypted alien personality is represented as a glassy ripple across a background of stars. Actually, it's more than a ripple—it's a wave-train; that is, a series of ripples several thousand miles long.

There are many such wave-trains, and they move through each other as transparently as the spreading rings from two rocks thrown into a pond.

Now the image in Frank's head shows a planet, none other than dear, fat, round Gaia Earth. An alien-bearing wave-train approaches it and then, upon nearing Earth's clamorous vibe, the ripples suddenly rear up, thicken, congeal and—out pops a Gray, two Grays, five of them, and a flying saucer.

"Oh, but you don't like the Gray shape," says Herman's voice in

FIGURE 27: Aliens Decrypting Themselves

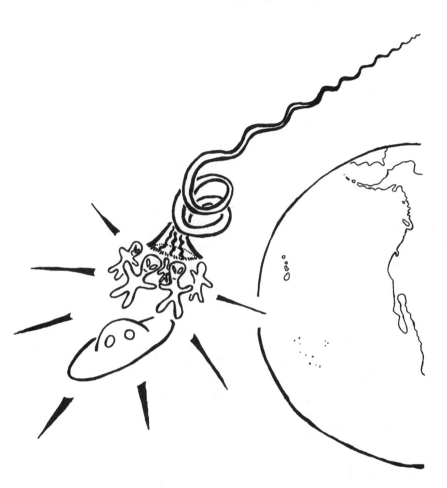

Frank's head, and the cartoon's saucer aliens change into—Frank fights back an involuntary spasm of disgust—five liver-colored starfish with big, red jelly eggs.

"Watch just one more time," says Herman, and replays whole sequence again. This time, Frank notices how the wave-pattern of the transmitted aliens first passes Earth, then curls back around and intersects itself. The moiré interference fringes of the waves make a higher-order pattern, and the higher-order pattern bends around again, intersecting itself at a yet-higher level, creating yet bigger peaks and valleys, and these in turn bend around again.

The mental movie is slowed down so much that Frank can make out level after level of this taffylike folding of the wave pattern upon itself. At the culminating moment, bumps of the rippled spacetime pinch off into little globs, like rapid water splashing as it flows across a shoal. The drops turn into five starfish and a flying saucer and now the ripples damp down, having converted their energy into something like mass.

"Why do you even have flying saucers," asks Frank, "if really you're just reconstituted higher-dimensional electromagnetic vibrations?"

"The saucer is a kind of interface," says Herman. "A holding loop that lets us stay stationary at a given space location without rushing on and on like radio waves. Also, of course, once we create these semblances of bodies for ourselves, it's comfortable to have a place to put them. And don't forget that we need a guest room for visitors like you."

Lightspeed Travel

"What's it like when you travel at the speed of light?" asks Frank.

"It's wonderful." And Herman gives Frank a mental taste of it.

As Frank and I later work out in conversation, the key thing about light-speed travel is that when you are going that fast, every place is *here* and every time is *now*. Put more dryly, the relativistic length contraction is such that, to a photon, the space ahead looks like a squashed sheet of no thickness. And, owing to the relativistic time dilation, from a photon's point of view the clocks of the world seem all to have come to a stop. At light-speed, distance shrinks to a point and time to an instant.

(Note by R.R.: This struck me as a fine paradigm for enlightenment, reminiscent of a koan sometimes attributed to the Zen monk Dogon:

"I say, 'I'm always here; it's always now; I'm always I.' And you say the very same thing. Here and now, I ask you this: in what respect are we different?")

Frank, who has a practical turn of mind, tries to get a handle on light-speed travel in terms of driving from the Bay area to Lake Tahoe in the Sierras. He puts it to Herman like this:

"Viewed as a whole, the drive from here to Tahoe is a fairly compact experience, because I drive at seventy-five miles per hour. If I drove at billions of miles per hour it would be an *extremely* compact experience. It would almost be like everything is here and now. But I don't get how you know when to stop. Like, say I'm driving a billion miles per hour to Tahoe and I suddenly holler out, 'Hey, let's stop at that place we just passed, the spot with the mini-golf and the go-carts!' But I'm a long way past that spot by the time I even can manage to say 'Let's stop.' And if I'm traveling at the speed of light, it's even worse. How do you stop in time, Herman?''

Herman gives a rather long and discursive answer to this question of how the aliens manage to decelerate from light-speed and stop on a dime. Apparently, cosmic space includes a number of higher dimensions beyond the four of space and time. And some of these higher dimensions are "compactified," that is, curled up like cinnamon sticks. If I'm a smart light-speed pattern, then I'm able to whip around, say, the curled-up fifth dimension so as to do an instant U-turn, which accounts for that folding-over thing Frank was seeing.

Mental guttersnipe that he is, Frank visualizes the U-turn in terms of a stripper swinging around a fire pole in a topless bar such as the Pink Poodle of Santa Clara, to which Frank makes a pilgrimage at least once a year.

Herman is puzzled by the analogy, so Frank tries again, imagining the light-speed trip and the sudden, discontinuous stop as happening in a single, extended moment of mental time, a peak experience comparable to that of a basketball player ''in the zone'' who leaps into the air, twists, slam-dunks the ball, floats down and lands, with the whole sequence feeling like a single, undivided slow-motion gesture.

This makes sense to Herman and he readjusts his representation of a light-speed experience and feeds it to Frank again. What an incredible sensation it is. You (the alien who travels hither from the great yon) have this *whooom* speed-of-light trip that lasts, strictly speaking, no subjective time at all (even though you *do* experience it!), and then you swoop down on some fresh world, and it's minutes, hours, years, centuries, millennia, eons later than when you started and you're correspondingly far away from home, but it's like no ''time'' elapsed during your trip, your trip is a radical discontinuity, a nonlinear spike, a shock front.

"It's not quite right to think of the trip as being instantaneous," corrects Herman. "Things do happen during the trip, only they all happen at once. We cross the signals of lots of other aliens, for instance. And there's the big cosmic-background signal as well. God."

LIFE IN THE UNIVERSE

The alien guest book outside time like Akahasic Records. But the Blukka built a ziggurat.

Wetware means gene. Herman's genes are like pentagons.

Shuggoths and sunspots are alive without wetware. Even the galaxy.

God is everywhere.

Meet Rudy at Jahva House 11:30 tomorrow. Ask for 50%, he'll agree on 20%.

The Alien Guest Book

"You encounter lots of other encrypted aliens on the way here?" asked Frank. "How many different groups have visited Earth, anyhow?"

"Not so very many," says Herman. "Perhaps a hundred thousand. Trillions have passed by, but most of them don't stop. Many of those other aliens are from the insides of stars or from cosmic black holes. A planetary wetware world like Earth means no more to them than a pebble would mean to you."

"You're telling me that there's been a hundred thousand alien visits to Earth?"

"There's something like a guest book. Whenever we stop somewhere we leave our memories for future visitors. These are stored as quantum-interference fringes in paratime. Now and then some of the less ecologically-correct visitors have taken the step of influencing humans to build certain terrestrial monuments or works of art. Most of us feel this is a bad thing to do. If everyone did it, your planet would be a shambles. But actually, here I am meddling very overtly by giving you all this information for your book. I suppose your *Saucer Wisdom* will be my saucer's recklessly garish 'guest-book signature.' "

Frank thinks of a book of "hobo signs" he saw as a boy, and wonders if the compulsion that drove a pharaoh to erect the pyramid of Giza might be simply a magnified fence-scrawl meaning something like "bad dog here" or "ask wife for handout" or "good eats if you do chores."

Catching Frank's thought, Herman beams him a guest-book vision of ancient Mexico. Thick, rank green vegetation. Hovering in the mid-distance is a saucer, shiny and metallic, but of a rich vermilion hue. The saucer is beaming yellow-green rays down at Mayans who labor on a great ziggurat. A closer look shows that the high priest has a blobby, purplish asymmetric alien resting on his back, with a tangle of rootlike feelers reaching in through the priest's neck to control his brain.

"A terrible crime," says Herman, and the vision ends. "Those purple aliens made the Mayan society collapse. We've seen the traces of those ill-shapen spine-riders before," says Herman. "We call them the Blukka. Their behavior is so beyond the pale that I'm sure they don't come from a wetware world."

Wetware Worlds and Otherwise

"So what *is* a wetware world?" asks Frank.

Herman explains that this is an environment, typically a planet, where each creature in the ecosystem has a compact program from which the creature's physical body is grown. In wetware worlds, the creatures reproduce by making small copies of their programs and growing new organisms from the copies. A wetware world is viewed as really inhabited by one kind of thing: the programs that grow its creatures.

In wetware worlds, one privileged form of information replication has managed to extrapolate and exfoliate itself so as to occupy every available environmental niche. Small and large variations in the programs are repeatedly tested out by the mechanisms of mutation and sex. Evolution inevitably arises. The species relationship-diagrams of wetware world have a characteristic branching structure. Every creature in a wetware world is "kissing cousin" to every other.

From the alien point of view, Earth is a wetware world with one basic kind of thing living on it: the versatile, self-replicating DNA molecule. All of Gaia's fauna and flora are close kin: Each of our species is DNA based, and each of our Terran organisms has much the same kind of metabolism—a tidy little molecular clockwork of hydrogen, oxygen, carbon, and light. The whole panoply of Earth's life could, in principle, evolve again from any individual representative.

FIGURE 28: Ladder Wetware, Sheet Wetware, and Block Wetware

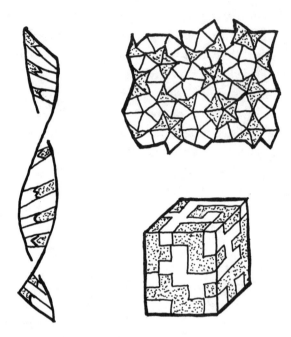

Just as every human language is quite different, being based on a number of arbitrary low-level code-symbol decisions, each wetware world's molecular biology is unique. In some of them, the master molecule is a double-stranded thing like DNA; the advantage of DNA being that it always has its spare copy ready. In other wetware worlds, the wetware code is in a two-dimensional sheet that is able to grow copies of itself like a rubber stamp printing an image. In still other worlds, the wetware code is a three-dimensional block that replicates itself by quantum resonance.

"Our wetware is in fact a two-dimensional quasi-crystal with five-fold symmetry," says Herman. "This makes it easier for us to breed different groupings together. Quite often, five distinct members of our species will combine their wetware to create one child. Each of us supplies, as it were, one leg of the adult."

Frank has difficulty in imagining a world that is *not* a wetware world,

FIGURE 29: The Shuggoth

so Herman shows him an example, a kind of mental travel movie of a world that he and his companions recently visited. Watching this memory tape feels like experiencing it directly, only from one of the starfish aliens' viewpoints.

In the tape, the saucer lands on a huge, misty plain. A big shambling thing—Frank thinks of it as a *shuggoth*—comes inching along. It is obscenely ungainly, a walking pile of odds and ends. The shuggoth emits a sour bellow, then wallows into a nest. In the nest are bits and pieces of shuggoth flesh, which the creature is cobbling together to make an offspring.

Other shuggoths appear; none of them is alike. The shuggoth world is a place in which globs reproduce themselves as wholes, almost like giant amoebas. ''They lack the unifying discipline of growing from a seed,'' Herman tells Frank. ''Such non-wetware creatures have no need to resemble each other at all. Things are tacked on, jerry-built, jury-rigged. There is no consistent upward evolution. It's like . . . like philosophy instead of like science.''

''Are there many non-wetware worlds?''

''Oh, yes,'' says Herman. ''All of the stars.''

Sunspots Are Alive

''For instance,'' Herman tells Frank, ''There is a race that lives inside your own Sun. It's rather easy for us to observe them, even from this distance. You call them sunspots.''

Herman shows Frank a view of a sunspot that's like a whirlpool, or rather like a pair of maelstroms lined by a subsurface vortex thread.

As Frank watches, the vortex wambles and splits off a child vortex like a tornado spawning extra funnels. ''Vortices are expert at non-wetware reproduction,'' says Herman. ''The children aren't grown from programmatic codes; they're cloned off whole. There's a sense in which a parent vortex cares for its children; it feeds rotational energy into them.''

''But are they intelligent?''

''Intelligence—to a wetware-based person, this means reducing the world's welter of information to small wetware-like codes. Aphorisms, poems, laws of nature, concise mental concepts. The sprawling non-wetware beings think in a different way. They emulate instead of summarizing. Look more closely at the sunspot's central axis. I've highlighted it for you.''

Now Frank sees a glowing purple space curve that runs along the twisting axis of the great sunspot. The curve is of a remarkable intricacy, like a tangled phone cord gone mad. The view zooms closer and Frank can see that what he thought was a single curve is in fact a braid of narrower curves. Odd wave forms travel up and down these woven lines, and each of the lines is thousands of miles long. And, of course, each of these individual lines is itself a braid of yet-finer structures. So much information!

FIGURE 30: Sunspot Vortex Threads

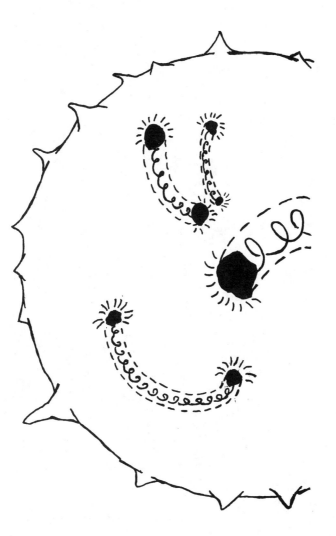

Frank remembers reading an article about how astronomers are always being surprised by the new behaviors of sunspots. Suddenly this
makes perfect sense.

"And of course there's non-wetware life much bigger than sunspots," says Herman. "A star as a whole is sentient, and so is your
planet Earth. But these minds are too slow and too vast for us to properly
encompass. And there are levels upon levels after that."

FIGURE 31: The Fractal-Line "Mind" of a Sunspot

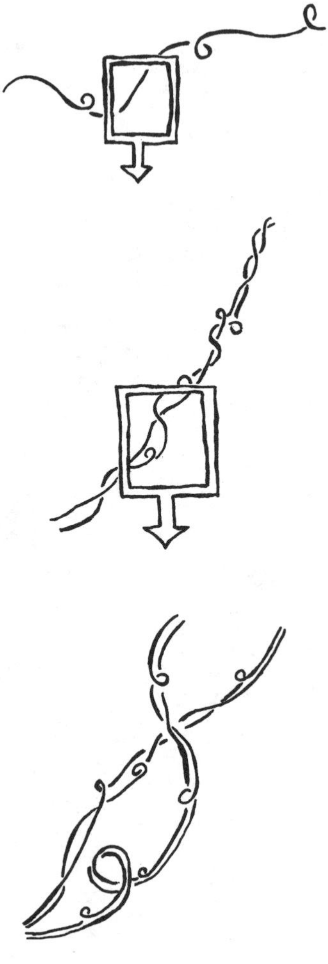

God

Herman gives Frank a sketchy description of a great chain of being that leads upward. Not only is there life on planets and inside stars, the planets and the stars *themselves* are alive. At a higher level, the galaxy itself is a living organism—at a timescale we have trouble grasping. The Milky Way galaxy, for instance, takes a quarter of a billion years to rotate around its center. Two hundred and fifty million years. And that's like *one day* for the galaxy.

"At the highest level," says Herman, "the cosmic-background radiation is the One Mind. The vibrations of God are always present, just like the radio waves that encrypt interstellar travelers such as me. God is always prepared to appear within you if only you open your . . . your heart."

After this ultimate bit of saucer wisdom, Frank finds himself back above his cabin. He's tired and hungry and ready to go home.

Temporal Attractor

"One more thing," says Herman.

"Nothing more," implores Frank, gazing longingly down at his little house beneath the redwoods. "I need some time to process all this."

"This will just take a few minutes. We're going to dart over to Santa Cruz and jump up to tomorrow so you can see your meeting with Rudy."

"What meeting?"

"Tomorrow you meet him at the Jahva House. You nail down your deal and get started on the book. *Saucer Wisdom*. We're counting on you two to do it. If you view some alternative versions of your negotiation with Rudy, then you can figure out the best approach."

So the next thing Frank knows, he's floating in a tiny, invisible saucer in the Jahva House, watching himself talking to me. The odd thing is that he sees sixty or seventy versions of the conversation in a row. No sooner is one version done, than the saucer loops him back to the start of a fresh one.

In some of the talks, Frank's demands enrage Rudy so much that he leaves, in others Frank asks for so little that Rudy agrees with condescending readiness. As Frank watches these variations on his future, it's like he's able to reach into them and manipulate them, moving the future timeline back and forth like a stick in a rapid, turbulent stream. Sometimes a little change has a big effect, other times it does nothing.

Slowly, the repeated versions stabilize and converge on a single, optimal strategy. Frank asks for 50 percent, keeps Rudy from leaving, then "settles" for 20, which is evidently the maximum that Rudy will pay.

"The future's not really fixed?" Frank asks Herman as the saucer speeds him back towards his cabin.

"It *is* fixed," says Herman. "You need to let go of your old ideas of cause and effect. All parts of time can affect all the other parts, and the single, real timeline is the holistic sum of all the back-and-forth effects. Good-bye for now.''

The saucer lowers Frank down into his desk chair and then the alien disk dissolves back into a complex of bright patterns that merge into the flickering of the three TVs.

A New Style of Ufology

In closing out our discussions of his alien notes, Frank went into a long rant against contemporary ufology and its obsession with government secrets.

"I mean, *please!*" he concluded. "Why concern yourself with paranoid fantasies about the *government,* for God's sake, when the simple truth of the aliens is all around you, as pervasive as the pollen in a May breeze? Do it yourself, and don't expect some lying moron of a politician to do it for you! Alien consciousness is yours for the taking."

I pressed him to tell me what he meant by this and he made the following comments. "At any given time it's entirely possible that an invisible alien might be right at your shoulder. Instead of denying this, or believing it but being frightened, why not accept it as true and rejoice in it! If I have an alien at my side, I can share in its wonderfully fresh view of things. I can see the world through new eyes. The more I see our world as an alien might see it, the happier I am.''

SEVEN

Notes on Future Biotechnology

By Tuesday, June 21, 1994, I had the first two chapters based on Frank's notes written up in much the same form that you see them here. I was excited about the stuff and I was cranking pretty hard, writing all day long every day, weekends and evenings as well. Audrey thought I was overdoing it, but she was pretty busy herself with her paintings. Her show at the Coffee Roasting was due to start on the first Saturday in July.

On the morning of Wednesday, June 22, 1994, Frank phoned me up and I was able to tell him that I was all done and ready for more notes.

"Bring the chapters down to my cabin so I can read them," he said.

I sort of wanted to say yes. But I remembered having made a big issue about not wanting to have Frank reading the book before it was done. And I remembered why. Frank would be too likely to give me a hard time about the details of what I'd written. For instance, to me it might seem funny and accurate to call Frank a "mental guttersnipe" for dragging Pink Poodle stripper gymnastics into a discussion of rolled-up higher dimensions—but an edgy loner like Mr. Shook would probably take that kind of remark as a mortal insult. Even more problematic, Frank might object to things that I didn't even suspect might be offensive.

"So what if he *were* to object?" you might say. "Big deal." The problem is, when someone criticizes something in one of my works in progress, I start doubting myself and changing things—and then changing them back—all the while expending huge amounts of energy on silent internal arguments with mental straw-man simulations of my critic. So I said no to Frank's request.

"I'm sorry, Frank, I don't want to. Remember when we made our deal at the Jahva House? You get your twenty percent, but you have to trust me while I'm writing the book."

"But what if you get something wrong? Don't you care?"

"We'll have one big correction session when I'm all done, and for sure I'll fix anything that's a mistake. But I'm worried that if I start going back and forth with you now, I might lose my momentum. I'm always in this state of anxiety when I'm writing. Like that the Muse might leave."

"I'm honored to be working with such the literary *artiste,*" said Frank sarcastically.

"Okay, let me put it another way. It's like I'm fucking and I don't want anyone to interrupt me before I come."

"What a way for a professor to talk! Well, all right, so you're not gonna show me the book till it's done. Are you writing it on your computer?"

"Yeah, I always use a word processor. It's a lot easier. Could you get some more notes for me? Try and ask about my other Great Work question. The future of robots and artificial intelligence. Something practical to counterbalance all that wild alien stuff, not that I don't love it."

"I could try and attract a saucer for a session today. I haven't been with the aliens since June eighth. I'd been hoping to spend the day with you, but since you're being such a prima donna about showing me what you've written, I might as well be a good collaborator and get you more material."

"Good man! Do it. And then I'll come down and see you tomorrow and you can give me the notes and explain them to me."

"All right. We can meet at Carlita's."

So on Thursday, June 23, 1994, I had lunch with Frank at Carlita's again. It was much the same as before. I had chicken soup; Frank had an enchilada platter; we both drank beer. Frank had brought in a sheaf of notes on future biotechnology. It seemed the aliens hadn't been interested in doing research on robots. We sat at Carlita's for a couple of hours, going over the biotechnology notes together. They were good.

When we finished with the new notes, our conversation drifted around for awhile.

And then I brought up a problem that was still bothering me a lot: Peggy Sung. "Should I be worried about this woman? Is she going to try and do something to me?"

"Well, as it happens," answered Frank, "I ran into her Sunday at the supermarket. In the flesh. She knows all about what we're doing, and she thinks she ought to be getting a cut of the book money too. Like she said on the beach. She wants ten percent."

"What!?!"

"Well, it can be argued that she invented the triple-feedback idea. And there's a sense in which triple feedback is the fundamental technology that's making *Saucer Wisdom* possible. It's like if you were using someone's camera for a photography book, you might have to pay them a licensing fee. I don't necessarily agree with her, but that's the way she looks at it. And I'm afraid she can tell the aliens not to come for me anymore. Basically it's either kill her or pay up. And if I killed her I'd get in too much trouble."

"You practically tried to kill her on the beach. Throwing that rock at her."

"Aw, I knew it was gonna miss. Do you think I'm a psycho or something?"

I sighed and looked out the window. There was a Pacific Bell truck slowly driving by. The driver was a dead ringer for Guster; he was staring at me, but he didn't wave. And sitting next to the driver was a longhair who looked a lot like Spun. The truck tooled past and was gone.

"Spun and Guster!" I exclaimed. "I think I just saw them again! Driving by in a Pac Bell truck! Guster was at the Pot of Gold phone booth across the street the first time I met you here, and then he and Spun were at Jahva House two weeks ago. Do they have something to do with Peggy Sung? Are they spying on me? For that matter—are *you* spying on me? Is that how you found out I was going to call our book *Saucer Wisdom*?"

"Don't turn all paranoid on me, Rudy," said Frank. "It's a small world up here in the mountains." He'd had several more beers while we were talking about the future of biotechnology. "I'm the one who should be paranoid, the way you won't show me your manuscript. I think you're way out of line on that. Why the hell shouldn't I know what you're writing about me? What if you're calling me a nut and an asshole?''

"If you want me to, I'll say it to your face, Frank. You're a nut and an asshole." I forced a laugh. "Christ. You people are driving me crazy. So now I'm gonna go home and work on the actual book. Give me a call next week, maybe Wednesday."

Despite the tension, the biotechnology notes were great fun.

CHIMERICAL BEASTS

Herman the starfish again. Robots are junk. Wavy nature. Dog that licks your floor, all tongue, no dog, brain of a worm, hyper mole guts, Big Tongue. Better than LuvSlug, it's alive. Sausalito Suzy Creamcheese likes it, her airhead me-show.

Biobot engineers can see DNA, high-level. Anti-sound bees glue motors shut, psychedelicize thy neighbor. Neatnik bird pecks trash. Super blowfish in pool.

A purple dog with orange spots, very cute. And then—look out for the compsognathii! Mean as dachshunds. Dona and Chu tame them with spaniel-brains and simulated evolution. Pet dinos at last! Roar! A pterodactyl pecks a cat and misses.

Machines Suck

Around noon on Wednesday, June 22, 1994, Frank starts up his three video cameras and TV sets. This time, it takes a little longer than usual to zero in on the right kind of alien-attracting fractal. The idea is to crank all of the cameras' sensitivities down to very low values and then to inch them up more or less in synch.

Eventually the organic-looking little feedback shapes are twitching around on the monitors and the screen thickens up. The effect reminds Frank of those pictures that turn into a three-dimensional image if you look at them with your eyes crossed. One instant he's looking at patterns on a screen, but then his focus somehow changes—and he's looking at a hovering bright-line saucer that moves slowly forward to engulf him.

The aliens hook into Frank's mind, and he recognizes the vibe. It's Herman the starfish again.

"Hello, Frank," says Herman. "We're here for another excursion with you. How's it going with your book called *Saucer Wisdom*?''

"Oh, fine, I guess, but Rudy won't show me what he's writing. He wanted me to ask you about the future of robots."

"Robots?" says Herman. "What a stupid question. It's strange how obsessed you humans are with your machines. You're so monkeylike, so *simian*. Fiddle fiddle. There's no reason why there should there be a clockwork machine that acts like a person or like an animal. The idea of making things out of machinery is childish; it's magical thinking. Like making goddesses out of clay. Or putting pictures of money in your wallet. Or pasting cardboard on your arms and thinking that'll help you fly. Look."

The saucer darts through some near-future scenes of clunky mechanical robots falling over, crashing into things, getting stuck in unproductive oscillations. One image shows a robot's metal-strut arm bumping a door frame over and over again, finally gouging out a hole in the frame and the wall, with the arm getting bent and battered as well; at the end of all the clumsy bashing, the robot totters through the ruined door and immediately falls down a flight of stairs.

"Robots are such a stupid idea," repeats Herman. "Piezoplastic will change the situation a little, but even soft robots are just machines. Biotechnology is much more important than anything humans can do with machines."

"All right," says Frank. "What does Rudy know? Let's do biotech instead of robots."

Biobot Inc.

Frank and the aliens skip into the future and zoom in on a start-up company called Biobot. They're in a flat building near the Bay in San Mateo. Frank and the saucer hover there to watch Biobot's techie founders making their pitch to the venture capitalists.

"Our first product will be called Big Tongue," says one of the techies, an Indian man standing in front of the room. His name is Saleem; he has inspired, liquid eyes. "Big Tongue will be positioned as a robot vacuum cleaner—although it is not a robot. It licks off your floor all the time, like an eager dog who slurps up a spill. Note that Big Tongue is *not* mechanical, it is bioengineered from a cow's tongue crossed with the nervous system of a flatworm and the digestive tract of a shrew." Saleem smiles, takes a drink of ice water. "Such a collage creature is called a *chimera*, after the Greek myth of a beast that was the combination of a lion, a goat, and a serpent."

"D'oh!" says one of the venture capitalists, a fat, yellowish man. "How in the world are you going to market this, Saleem?"

"I'm glad you ask," says Saleem. "Big Tongue has four selling points, which we are summarizing as *Big Tongue licks clean*, *Big Tongue cares*, *Big Tongue keeps mum*, and *Big Tongue licks you*." A sample ad appears on a big flat "Abbott wafer" screen behind Saleem, prominently featuring the four slogans.

"The first point is the obvious one," says Saleem, pointing to the screen. "Big Tongue will be keeping your floor impeccably proper. *Big Tongue licks clean*. No artificial substance can match the flexibility and roughness of a real tongue.

"The second selling point is that Big Tongue is organic. A mechanical device is something that many people fear and mistrust—with good reason! Machines—and I include computers and piezoplastic in this category—require upkeep, they malfunction, and—this is the most significant—they are an entirely different order of being. Big Tongue is our kin, Big Tongue is based on DNA, *Big Tongue cares*. It participates, if you will, in the Gaian morphic resonance.

"The third point is that Big Tongue is not a burden to have around. Unlike with a maid or a servant, you don't feel guilty towards it. Big Tongue is organic, but at a scale so low that it does not carry the responsibilities of a pet, not even of a goldfish. Nobody need have a bad conscience towards a cow's tongue. *Big Tongue keeps mum*.

"A final selling point is that Big Tongue is readily amenable for use in personal grooming. Think how pleasant it is to be a kitten washed by a mother cat. Big Tongue will lick the user wherever he or she likes. *Big Tongue licks you*."

A short-haired man in a silvery suit raises his hand. "This personal massage angle, Saleem. We're not going to take a low road on that, are we?"

"Oh, no, no, no," says Saleem. "Leave the low road to people's low imagination. We mention only that Big Tongue is naturally antiseptic, and is always clean enough to be used on one's own body. And we will perhaps remark that Big Tongue might be playing the role of a sponge in the shower to wash off the areas that are difficult to reach."

Big Tongue vs. Lickin' LuvSlug

Big Tongue hits the market and Frank gets the aliens to zoom in on a user, a Suzanna Quarkaese in Sausalito, just north of San Francisco. She's in the kitchen talking with her friend Benny Gay. On the floor are a Big Tongue and a piezoplastic LuvSlug.

"You bought both of them, Suzanna?" exclaims Benny. "You are such the rich bitch."

"I couldn't decide," said Suzanna. She's actually not bitchy-seeming at all; she has warm features, long brown hair, and a generous mouth made bright with yellow lipstick. "I think they're both super-exciting. The new-model LuvSlugs clean the floor too, you know. It's a just-in-time feature so that they can compete with Big Tongue. They're called Lickin' LuvSlugs. I've been testing them out all morning. Can you run my dragonfly, Benny?"

"You want to shoot this?"

"Of course, that's why I asked you over, duh. I'm going to make a little news feature comparing the Big Tongue and the Lickin' LuvSlug—for the Suzanna Quarkaese Me-Show. For my hundreds of fans. Once I had four hundred and sixty-three viewers!"

"Ooooh-kay," says Benny. He puts a piezoplastic uvvy patch the size of a silver dollar onto the back of his neck, and a little plastic dragonfly hops up off its perch on a kitchen shelf and begins flying around.

"Ta da!" says Suzanna. "It's me again, Suzanna Quarkaese. Guess what I just bought? A Big Tongue and a Lickin' LuvSlug. One's meat and one's smart plastic. Watch along as we compare and contrast!"

Suzanna opens the fridge, gets out a pint of nasty, moldy old yogurt and dumps it on the floor in two piles, one near each of the two cleaning products. The Big Tongue is pink and rough, very like a cow's tongue. The LuvSlug is gray and iridescent. Each is about a foot and a half long, and some four inches in diameter. Neither one is particularly pleasant to look at.

When the yogurt plops onto the floor, it's the Big Tongue who notices first. It must have something like a sense of smell, for no sooner do the curds hit the linoleum than the Big Tongue writhes over to the spot and starts licking. There seem to be fissures in the Tongue, for the yogurt disappears right into it. As it continues to lick, Big Tongue begins giving off a low, purring sound, like a mother cat tending a kitten.

"Wake up, Lickin' LuvSlug," says Suzanna and nudges it into its patch of yogurt with her foot. Once the LuvSlug feels the moisture it begins rapidly scouring the floor, absorbing the stuff into its body just like the Big Tongue, but soundlessly. The Lickin' LuvSlug finishes off its spill and begins casting about for more to clean. It bumps up against the Big Tongue, which is still ruminatively licking its area of floor. The two dumb sausage creatures bump and nuzzle each other. Big Tongue's purring takes on a questioning tone.

"Get a close-up of that, Benny," says Suzanna. "It's too absurd."

"Throw something else on the floor," suggests Benny.

So Suzanna tests the cleaning-creatures with butter, ketchup, flour, a box of cereal, and a big spill of water. The Big Tongue is consistently more responsive and thorough, but the Lickin' LuvSlug moves faster. After the tests, both of the slug things are somewhat distended from all the stuff they've absorbed.

"I bet you're wondering how I empty them," says Suzanna. "First the bad news." She picks up the Lickin' LuvSlug and squeezes it over the sink. A thin, milky gruel leaks out of one end of the LuvSlug and runs down the drain.

"Gross!" exclaims Suzanna. "And *don't* forget to rinse." She runs water over the Lickin' LuvSlug until the stream from its bottom end runs clear.

"And now the good news!" Suzanna makes a lo-and-behold gesture with both hands, pointing at the Big Tongue, which is sitting on the floor, quietly shivering. "Big Tongue metabolizes its meal away! Now, let's see how they work on carpeting and woodwork."

Frank's getting a little bored with this, so he gets the aliens to skip ten minutes forward, to the inevitable shower scene.

Suzanna is in her shower stall, nude and attractive, and she's got the Big Tongue in one hand and the Lickin' LuvSlug in the other. Benny Gay is sitting in a chair in the bedroom, flying the dragonfly around the bathroom.

"Am I the lucky girl or what?" says Suzanna, holding up the two slug things, one pink and one gray. The Big Tongue reaches over and slurps against her cheek, while the Lickin' LuvSlug rubs her arm. Suzanna closes her eyes and presses the two soft, writhing objects against her neck, then opens her eyes very wide.

"Oooo! Big Tongue's *totally* the best. I guess there's nothing like DNA. 'Scuse me!" The Lickin' LuvSlug drops to the tile floor outside the shower and Suzanna disappears into the steam. Loud purring. Benny flies the dragonfly out and away.

Wetware Engineering

Frank is curious about how the Biobot engineers designed the Big Tongue, so the saucer zips over to their labs. A few hundred Big Tongues—and what look like other Biobot products—are growing in vats of viscous blue amniotic fluid. In a workshop at the end are two scientists named Gene and Jean. Gene is a young man with burr-cut hair and a receding hairline. Jean is a rather fat woman in her thirties.

Gene and Jean have uvvy patches on their necks; this gives them radiotelepathic contact with each other, and also links them into a virtual-reality simulation the big Biobot supersluggie computer is running. The aliens give Frank a feed of what the two wetware engineers are looking at: a spiraling double helix of spheres that represents the DNA of a bumblebee.

"I hate my neighbors," Gene is thinking to Jean. "They make noise all the time. They have lots of electric machines. Pool pumps, leaf blowers, loud radios. They're morons. Today I want to invent something to stop them. Let's chunk up this bumblebee's DNA, see if there's anything there to work with."

"When did you get this DNA model?" Jean asks. "I haven't seen this one before."

"Oh, I downloaded it from the Department of Agriculture," says Gene. "They're using the scanning tunneling microscope to map the genes of all kinds of useful animals. Of course, they don't have your techniques, so they can't do all that much with it. Do your thing, Jean."

"All right," says Jean, and a sylphlike silver version of her appears in the shared mental cyberspace where the DNA molecule is floating. The winged Jean sylph darts up and down the long double helix, and the molecule is replaced by a complicated-looking machine of springs, gears, and levers. "This is abstraction-level two," says Jean.

"Run it," says Gene.

Jean sets the big virtual DNA machine into motion, and its components shuttle this way and that, weaving tapestries that scroll out of slots all

**FIGURE 32: Suzanna in the Shower with Big Tongue
and Lickin' LuvSlug**

over the long machine. The glittering little images show various pictures of pieces of a bumblebee: legs, feelers, compound eyes. To make the vision all the more wondrous, many of the pictures are in fact animated; the legs, for instance, are repeatedly flexing over their entire range of motion.

"Okay, I think I see where I want to do my programming," says Gene. "Can you raise us on up to abstraction-level four, Jean?"

"Meta meta meta," says Jean, and now the DNA takes on the appearance of a short spindle with distorted pieces of a bumblebee set into the shape. It reminds Frank of an image he once saw of a human brain with the corresponding body parts drawn on top of it: a big tongue all across the cortex, bulging eyes on the side, tiny hands and feet. But it's not just the physical parts of the bumblebee that are present in this level-four view; the behaviors of the bumblebee are there too, represented as little film loops—wing-flapping is over here, flower-seeking over there, mating on this side, and so on.

"Yes," says Gene. "Now, what if I take this little ear membrane and make it bigger. No, better, I'll cut and paste it thirty times. Snip snip snip glue glue glue, and the bumblebee has thirty ears. Wavy. And now the behavior: I'm going to make the bumblebee hate sound. But only a certain kind of sound. Electrical motors. Drop the view of the ear DNA down to level three, would you Jean?"

"Okay," says Jean, and a graph of the ear's stimulus-response curve appears; Gene fiddles with it to put a big hump in at the sixty-cycle-per-second frequency characteristic of electric motors.

Now Jean has a thought. "You haven't thought about how a bumblebee is supposed to attack an electric motor, Gene. He can't just fly in there and get squashed. We need to give him something to use to jam the motor."

"Glue?" suggests Gene.

"Yes!" says Jean. "Remember the barnacle DNA we were looking at? And how it grows the glue glands the barnacle uses to attach itself to a rock? I'll call up some barnacle DNA right now."

A new level-four DNA representation appears and Jean and Gene collage a barnacle's glue glands into their sound-hating bumblebee.

"I just thought of something else," chortles Gene. "Let's put Colorado River toad venom into its stinger. That stuff is supposed to be incredibly disorienting. This way if anyone tries to stop my bees, they'll get stung and the venom will flip them out so bad that later they won't even remember what happened. It wears off in about five minutes, I understand."

"I like it," says Jean with a soft chuckle. "My neighbors have a pool pump too. Let's just pull down your toad DNA from the Web." Jean's virtual shape makes some quick, efficient gestures, and the toad DNA appears.

"Of course, this isn't exactly a product that Biobot is going to be able to advertise," says Gene as he finishes editing his chimera's genes.

"Oh, maybe Saleem will think of an innocuous way to position it," says Jean. "He's good at that. You're done editing? All right, I'll have the synthesizer fabricate two hundred of these gene strands and then we'll put them into some living cells and grow the new—what shall we call them?"

"ShushBees?" suggests Gene.

The ShushBees, the Neatnik Bird, and the FlushFish

Frank and the aliens zip forward a few weeks to find Jean and Saleem knocking on Gene's door late one Sunday morning. Saleem has a bag of bagels, and Jean holds a little basket with a domed woven lid: a miniature beehive.

Gene's house is a low, beige bungalow on a hillside. And just down the slope from Gene is a sprawling long, yellow house with a swimming pool. These are Gene's noisy neighbors. Their pool pump is a shrill drone. Country music plays from loudspeakers by their pool. A blank-faced neighbor woman comes out of the yellow house and starts up a leaf blower. The mechanical shriek drowns out all other sounds; the big wonderful world shrinks to shrilling point.

The three Biobot scientists repair to Gene's rickety deck, and Jean removes the lid from the little hive. Nine identically-plump ShushBees stare up with their compound eyes. They're nestled in the hive head to tail like a scarab bracelet. The leaf blower screeches on, numbingly loud.

Moving in unison like aerobatic biplanes, the nine fat gene-tweaked bumblebees rise up and circle in the air. It seems as if the bees can sense the transparent little lens of the saucer that hangs above the scientists; they fly repeatedly at the saucer, only to bounce harmlessly off it. The

FIGURE 33: ShushBees vs. the Leaf-Blowing Neighbor

attack of the bees is a bit unsettling to Frank, though the aliens seem to enjoy it. Quickly the bees give up on the slippery saucer and buzz down the hill towards the noisy neighbors.

"Bagel, anyone?" says Saleem.

There's a sudden bellow from down the hill, followed by a clatter and a wild cackling. And then, blessedly, the leaf blower stops.

"One fish, two fish, red fish, blue fish," the neighbor lady is hollering over the country music and the pool pump's whine. Then the music stops. And then the pool pump. It's a lovely, quiet Sunday morning in California. The air is cool as water, the sunlight's bright and full of fun.

Jean sets a tiny saucer of honey inside the hive; six ShushBees fly back to Gene's deck; apparently three of their number were squashed by the neighbor, or by the machinery. The survivors crawl inside their hive for breakfast.

Meanwhile the neighbor lady's raving voice mutates into song. "Oh, what a beautiful morning!" she roars. "Oh, what a beautiful day! Fuck working! Wooo-ork? I'm gonna to take a swim and just lay out and relax. Yee haw!" Great splashing, more singing, and then a profound yawn as she stretches out on her chaise longue for a nap. It's a calm, calm Sunday.

"*BeeLab*," exclaims Saleem. "That's the ticket. An experimental product with no explicit purpose. '*Learn more about our friend the bumblebee with the Biobot BeeLab. Each BeeLab includes nine ShushBees and three easily cleaned electrical buzzers. The genetically engineered ShushBees have been designed to home in on sound sources and to jam them with a gluelike marker substance. Warning, due to an unavoidable side effect, the stings of the BeeLab ShushBees are causing a short-lived psychotomimetic effect. Do not attack the bees! Wear protective clothing. Do not use around unprotected individuals. Do not use in the vicinity of unpleasant noise sources belonging to other people. Biobot Inc. assumes no responsibility for improper or irresponsible use.*' "

"He's never going to get away with that," says Frank to the aliens, and he's right. They saucer jumps a few months into the future, and Frank finds that the Biobot BeeLab never makes it to market. While Jean is testing it, a BeeLab ShushBee stings a leaf-blowing gardener, who then gets his leg broken by a passing car. Despite the toad venom, the gardener is able to remember that the ShushBee came from Jean's balcony, and Biobot ends up having to settle with him out of court.

So the ShushBee project dies, but the silence-obsessed Gene has the idea of making Neatnik Birds and FlushFish to replace leaf blowers and pool pumps.

Jumping a bit further into the future at Gene's house, the saucer is able to see a neighborhood at peace. Gene has given his downhill neighbors a Neatnik Bird, a FlushFish and a set of new uvvy receivers so that they can silently listen to their music.

The Neatnik Bird obsessively hops around the neighbors' yard, eternally gathering twigs, leaves, and trash for an ever-growing nest mound on top of their compost heap. It's based on a miniaturized ostrich with

FIGURE 34: Neatnik Bird

its nesting genes tweaked so that it always thinks it's about to lay an egg. The FlushFish is combination of a miniature whale and a puffer fish. It floats in a corner of the neighbors' pool sucking in water, filtering the liquid with its baleen, and spewing the water back out through a clever little blowhole.

Thanks to Biobot, the world becomes more and more a place of ease. Noisy machines give way to quiet animals.

Pet Dinosaurs

Frank and the aliens skip forward through time, keeping an eye on the UV broadcasts. Early in the twenty-second century, the proprietary Biobot DNA editing technology leaks out to the world at large. And pretty soon there's a pet-construction kit called Phido.

The saucer hovers at the shoulder of a twelve-year-old black boy named Jimi. Jimi's dad is well-off, and Jimi has an uvvy link to a fat sluggie processor that's been equipped with the Phido design kit. Jimi's sister Sara is sitting next to him, wearing an uvvy and kibitzing.

"They say a pit bull be a manly dog," says Jimi, picking an image of a sturdy, bulldoglike beast.

"I want him to be purple," says Sara, shoving in and changing the image's color before Jimi can object.

"That's juicy," says Jimi. "But, Sara, I'm thinkin' maybe a pit bull be too nasty. I want a dog I can wrestle with and not get bit."

"Give him a beagle's head," suggests Sara. "Beagles are gentle."

They try that, but the dog doesn't look right. "He's so ugly it hurts," says Jimi. "Let's work out with these Phido tools, Sara. We can *evolve* this pooch."

Jimi asks the software to do a Mutate step, and the virtual display shows nine possible variations on the purple pit bull–beagle blend. One of them looks sleeker than the others, so Sara and Jimi pick that one, then generate nine further mutations of it, pick the best, mutate off nine more variations, and so on. Along the way they go for a longer, fluffier tail. And some orange spots appear amidst the purple fur.

Once Jimi and Sara are done they send off the Grow command to a Phido-compatible incubatorium called the Daisy Hill Puppy Farm. At the farm, the custom DNA molecule will be assembled, implanted in a live cell—just about any kind of cell at all will do—and the cell will be goosed and cajoled into acting like a fertilized egg that can be grown to maturity in an amniotic bath. Of course, Dad has to okay the substantial billing charge.

Frank and the aliens are there when, a few months later, the Daisy Hill Puppy Farm truck arrives with the new dog. The custom hound is quite attractive. And very friendly. The upbeat outcome is a little boring for Frank; he and the aliens circle around looking for a Phido user who's doing something more transgressive and weird.

They light on a shy, skinny girl named Dona, aged eighteen. She has buckteeth and a big laugh; she's obsessed with dinosaurs. Her room is full of models and images of dinos. Her pal Wyla is there with her. Cute, blond Wyla isn't a social reject like Dona; nevertheless, she considers Dona her best friend.

"I'm figuring out how to use the Phido kit at the deepest level," says Dona. "So I can make dinosaurs."

"Isn't Phido just for making dogs and cats?" asks Wyla.

"If you read way down into the help files and the source code, it turns out you can make any kind of animal with it," says Dona. "It's not too well-documented. As if anything ever is. You're allowed to plug in a function pointer to an arbitrary user-supplied DNA code oracle. And, get this Wyla, I just downloaded a compsognathus DNA map from the Smithsonian Institute."

"Aren't they pretty vicious?" asks Wyla.

"Well, *yeah*," says Dona. "But I'm going to splice in part of a dachshund's brain so that maybe it'll act like a pet. Dachshund seems to fit, somehow. But, of course it'll still look just like a compsognathus. All green and scaly."

"I don't like scaly things," says Wyla. "I like a furry thing I can pat."

"To each her own, kiddo," says Dona, laughing.

Two months later, Frank and the aliens find Dona out in her backyard with a little flock of six compsognathii. They're green, little beasts the size of chickens. They have long necks and alert toothy heads. They jump up in the air a lot. Dona sets out a bowl of fresh chicken livers and the compsognathii are all over it, squealing, gobbling, and biting each other. Dona is wearing boots, glasses, heavy pants, a leather jacket, a tightly wrapped scarf, leather gloves, and a snug-fitting wool cap pulled down low over her ears. Nothing's exposed except for the tip of her nose.

In the background Wyla lets herself in through a gate in the yard's high fence. She has a boy in tow. He has dirty black hair and glasses.

"Hey Dona, this is Chu. I told him about your pets."

"Be careful," calls Dona. "They bite big-time."

"They are *so* lifty!" exclaims Chu. He steps forward and one of the compsognathii leaps at him. Chu gets his arm up in time to deflect the

FIGURE 35: Dona and the Compsognathii

small dinosaur from biting his neck, but even so, it makes a nasty scratch on his arm.

"Get out!" cries Dona. "It's too dangerous."

"We'll wait in the house," calls Chu. "I'll put some skin-tape on my cut."

Later in the house, Frank and the aliens hover there, watching Chu and Dona talk. Wyla's not around anymore. She's brought a boy for Dona, and now she's gotten out of the way. Frank can see out the window to the backyard where the compsognathii squawk and tussle. They're thoroughly unpleasant little monsters.

"They're not going to work out," says Dona sadly. "I'm going to have to kill them. I'll get them stuffed, maybe. They were expensive! I was greedy to get a whole half-dozen of them. Maybe I can get some money for them after they're mounted."

"It's a shame," says Chu. "But, yeah, I know a place that would buy stuffed compsognathii. It's a bar in the city. I'd like to work with you on the next round. I have some money too. There's gotta be a way to use the Phido program's Evolve feature to keep testing the *behavior* of the simulated mutants—instead of just looking at their *bodies*. Of course, starting out with a placid, herbivorous dinosaur would make things a lot easier. A triceratops or a brontosaurus."

"Yeah," says Dona. "But the carnivorous ones are so much more . . ."

"I know," says Chu. "I know. I've never met a girl like you, Dona."

The two dino geeks stare into each other's eyes, draw closer, kiss.

"And—don't use *dachshunds*!" says Chu a little later. "Whoa. Yap yap yap, bite bite bite. What were you thinking, Dona?"

"I—I guess I picked dachshunds because they're the dogs who act the most like dinosaurs!" laughs Dona. Her cheeks are flushed and her eyes are sparkling. She looks pretty. Chu kisses her again.

Frank and the aliens skip a year forward to find Chu and Dona with a full menagerie of dinos, all of them tame as can be. There are compsognathii, some friendly little triceratops, and a mud wallow full of sleek, green collie-sized brontosaurs. Chu has a special pet velociraptor, and Dona has a T. rex so gentle that she that can feed it raw meat right out of her hand. All the dinosaurs have the benevolently fuzzy brains of spaniels.

Needless to say, the tame dino pets are everywhere before long. A hearty roaring rises across the land. Perhaps the most striking of all are the pet pteranodons and rhamphorhynchii. These leathery-winged pterodactyls are gene-tailored so as not to exceed a wingspan of three feet.

Frank and the saucer spend a half hour following around a rham-

FIGURE 36: A Rhamphorhynchus Chasing a Cat

phorhynchus who repeatedly tries to peck a cat. Dog-brained as it is, the rhamphorhynchus always misses, and follows each miss with raucous I-coulda-had-ya squawking.

TWEAKED PLANTS

Amparo and Jose get a Giant Beanstalk cornfield with knives. It's good at first, then it's sad. Rusty rattling. The dead deer.

Try again. Garlic the size of pumpkins, strawberries like red devil heads. Yes. Señor Pepita gives them a gourd seed in a pizza-box. Mexican refugees live in the gourds.

All sorts of pods to live in. The Heironymous Bosch House made of four seed-cases. Plumbing and wiring included!

The Green Ball in the surf. Kelp bubbles undersea. Into space. Domes on the Moon.

Knifeplants

"Once you get started with wetware engineering, there's no stopping," Herman tells Frank. "We use it a lot on our home world ♦. So far we've only looked at examples of things humans will do to your 'animals.' Now let's see what you'll do to your 'plants.' "

In the telepathic mind-link, the quotation marks look mauve and wiggly. Kind of hanging there in the air. "Why the quotes?" asks Frank.

"On ♦ we don't have your egotistical plant-animal distinction," answers Herman. "All of our life forms can eat light a little bit—like your plants; and all of us can move—like your animals. And we don't have the brain/no-brain distinction, either. None of us has a 'brain.' " The snotty mauve quotes again.

"So how do you *think*?"

"With my whole body. Each of my cells is polyfunctional. Multi-purpose. You might say that my whole body is a 'brain.' Except that it's also a 'muscle,' a 'stomach,' an 'eye,' a 'testicle,' a 'tube-foot,' an 'ovary,' a—"

"I get the picture, Herman."

The saucer homes in on a farm in Gilroy. A young farming couple named Jose and Amparo Gutierrez are sitting on their porch staring off into space. The saucer flies around their heads, and Frank can see that they're wearing uvvy patches on their necks. The aliens hook Frank into the radiotelepathic virtual reality that Jose and Amparo are sharing.

It turns out they're looking at a seed catalog from a biotechnology company called Giant's Beanstalk. The catalog has the form of a simulated farm with plants that you can order. "Look over here, Amparo," Jose is saying. "Knifeplants! I wonder if we can grow these." The two of them are mentally wandering aroung the catalog farm.

There's footsteps on the porch of the virtual Giant's Beanstalk farmhouse and a simulated person comes down the steps, an old-timer in overalls who looks—right now, anyway—Mexican. He's a simmie who functions as the Giant's Beanstalk catalog agent. *"Buenos días,"* he says, grinning. "Call me Señor Pepita." Which is the perfect name, what with *pepita* being Spanish for *seed*. "I will tell you about knifeplants if you like, my amigos."

"Gracias," says Jose.

"Knifeplants require a soil that is extremely rich in iron, zinc, aluminum, and other metals," continues Señor Pepita. His voice is frank and cozy, with a bit of dust in it. He has a big mustache. "Does any of your land happen to be lying over landfill?"

"Yes, yes," says Amparo. "All of it. That's how we could afford it. When we plow we're always turning up old cans and bottles, rotten wood and nails, sometimes even—"

"Any sheetrock?" asks the Señor Pepita simmie.

"I don't know 'bout that," says Amparo.

"He means the crumbly white stuff," says Jose. "That's sheetrock. Yeah, we get plenty of it, Señor Pepita. A lot of construction waste ended up in this landfill, I think."

"Well, that's excellent," says the agent. "Because knifeplants need gypsum, and that's what sheetrock mostly is. I'd say you ought to get into knifeplants big-time, Mr. and Mrs. Gutierrez."

So they get a loan, place the order, and the Giant's Beanstalk knifeplant seeds arrive. They're pointy-yellow with a streak of gray. Frank and the aliens skim through the summer, catching about a frame a day, watching the growth in speeded-up time.

The knifeplants shoot up like cornstalks, straight and tall with big, floppy leaves. Their roots fan out very widely under the ground, searching out their diverse nutrients. The knifeplants require a huge amount of water, which brings in lots of weeds. Jose and Amparo are out in the knifeplant field every day from dawn till dusk.

Little green ears appear on the stalks in midsummer, three to five ears per plant. Jose and Amparo open one up. Within the enfolding leaves is a pale fruit, a tasteless, pithy thing like a unripe banana. Jose pushes the meat of the fruit away, and in its center is a gleaming metallic membrane, fragile enough to tear.

FIGURE 37: Jose and Amparo with Knifeplants

By late summer, the slender fruits are firmer and their flesh has filled with little kernels. The shining blades in the centers are tough and re-silient.

In the fall, Jose and Amparo go out and harvest the crop. Each ear has a woody stock at its base. The dry husks pull away easily, and the ripe, seed-filled fruits fall free. What remains is—a knife, a shining blade growing out of a strong, round handle. Some of the knives are only a few inches long, others are up to two feet.

Jose and Amparo set to work polishing the knives and cutting grips into the handles. The blades are wonderfully sharp and strong; since the metal has grown like a crystal, it has fantastic integrity. And the blade blends inseparably into the woodlike grip. This is only the second year that knifeplants have existed, so Jose and Amparo get a good price for their crop.

The Giga Gourd

The saucer jumps three years into the future, and the knifeplants have taken over Jose and Amparo's farm. The seeds have sprouted all over the place. Unfortunately, there's not enough metals left in the soil anymore, and the new crops produce knives that are weak and brittle, hardly worth bringing to market. Jose and Amparo are no longer careful about harvesting every last knife. Here and there dried stalks rustle, with rusting knives clanking on them. It's a melancholy sight, and it's dangerous to walk around in the fields.

Amparo is pregnant and she's worried about raising a child amidst all these knives. The transparent little lens of the saucer follows Jose as he walks his property, deep in thought. Jose finds a slashed-up deer lying by one of the knifeplants. The animal's furry, weather-rotted body is like a wet, dirty carpet. More knifeplants have taken root in the enriched soil beneath the flesh. "This is no farming," mutters Jose. "This is no way."

The saucer skips further along Jose and Amparo's timeline. The next year, they go back to growing the traditional Gilroy crop of garlic—and it's hell keeping the knifeplants from coming back into the fields. In fact, by late May, it seems like a losing battle. They hook into the Giant's Beanstalk catalog and ask Señor Pepita for help. Señor Pepita suggests they try his new wetware-engineered Supergarlic.

The Supergarlic has cloves the size of soccer balls, and it's a savagely territorial plant, easily able to overcome the encroaching knifeplants. Though no ordinary shopper would buy such grotesque cloves, the local powdered-garlic factory loves them.

Of course, soon the neighboring farmers are copying Jose and Amparo, and the price of Supergarlic begins to drop. Amparo, pregnant again, is enthralled with the idea of giant food, and at Señor Pepita's suggestion, Jose puts in an acre of Giant's Beanstalk's new Devilberries.

The Devilberries are strawberry plants that have been gene-tweaked into gigantism. The berries are the size of human heads; the plants' leaves are as big as flags. A drawback is that a Devilberry has a shelf life of only two or three days. Also, you need to put something under the berries in the field if you don't want them to have an ugly, muddy patch on one side. Also, slugs are crazy about them. Amparo uses a flock of geese to keep the slugs away, she keeps fresh straw under each berry, and she ferries the berries to market at just the right time. Nobody else does it so well as she. Amparo's Devilberries become known far and wide.

Amparo also finds a use for the huge, coarse Devilberry leaves: She dries them in the sun, knocks out the leaf material, and uses the network of dried leaf-veins as a fine-meshed fabric, good for packing the ripe berries.

Jose and Amparo have done very well in testing and popularizing the Giant's Beanstalk products they've tried: The knifeplants, the Supergarlic, and the Devilberries—not to mention the King Kong carrots and the Pontoon peppers. Upon Señor Pepita's recommendation, Giant's Beanstalk, Inc. now has a special relationship with Jose and Amparo Gutierrez. Theirs is an official research farm! Amparo's taste for gargantuan plants is nowhere near satisfied, so Amparo gets Señor Pepita to let her try out a brand-new super plant: the Giga Gourd.

The Giga Gourd seed comes in a big, flat cardboard box like an extra-grande pizza. Amparo plants it right next to their house so that she can baby it. Señor Pepita warns that the Giga Gourd requires an enormous amount of water, so Jose runs a pipe to drip right onto the Giga Gourd's bed.

That summer, the sprawling vines of the Giga Gourd crawl up the side of Amparo and Jose's house and completely cover the roof with awning-sized leaves. It's a good shield against the baking summer sun; inside the house it's cool and green. The Giga Gourd flowers are huge yellow parasols, so overpoweringly sweet that wasps and bees come by the thousands. Frank and the aliens buzz around in the bees' midst for awhile; the aliens really enjoy bugging insects. Meanwhile, Amparo makes window screens out of Devilberry leaves.

Once a dozen of the Giga Gourd flowers have set seed, Jose takes to picking off the other blossoms and drying them in the backyard. How many giant gourds is he going to need, after all? Nobody has any clear idea of a use for them. The dried flowers form tight, nearly transparent, yellow domes the size of one-man tents.

As the summer wears on, the Giga Gourd needs more and more water, a steady hydrantlike flow. Señor Pepita arranges to pay part of Jose and Amparo's water bill—so curious is Giant's Beanstalk, Inc. to see how the gourds turn out.

The twelve gourds swell to unprecedented sizes—as big as stoves, as refrigerators, as cars, as trucks—and finally they are as big as houses, or even bigger. Each of the gigundo Giga Gourds is patterned in a different arrangement of green, white, and yellow stripes. Some are round, some are long and thin, some are round with thin, crooked necks. People come from all over to look at them.

As it happens, there's a war going on in Mexico that summer, and *thirty-one* of Jose and Amparo's refugee relatives arrive in late September. Where to house them? In the gourds!

The gourds have gotten nicely dry in the autumn sun. Each of them has a hard outer rind, a two-foot-thick pithy hull, and a few score pizza-sized seeds within. Willingly, the refugees set to work fashioning the gourds into dwellings. There's no shortage of carving knives at Amparo and Jose's!

Each house has a swinging gourd-skin door and a few gourd-flower skylights. Boulders and dirt ramps keep the roly-poly homes from tumbling over. Jose uses a posthole digger to put in some external pit latrines.

Of course, the dwellings aren't up to code, but the city of Gilroy classifies them as temporary farmworker housing, and they're allowed to stay in place at least through the winter. The fanciful little encampment becomes known as the Casas Gordas.

Grown Homes

Frank and the aliens dart on into the future, watching the UV for news about the Casas Gordas. Bingo—the next summer, an architect-developer named Uli Lasser sees the bright fat houses, loves them, and goes into a huddle with Giant's Beanstalk, Inc. But the Beanstalk guys are too food-oriented; they just can't think housing. So Uli finds some venture capital and lures away two of the Giant's Beanstalk wetware engineers.

FIGURE 38: Casas Gordas

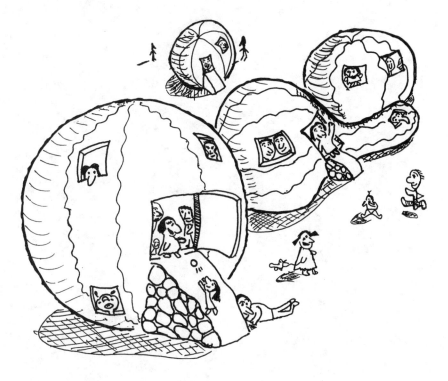

The UV announcement of his new company, Grown Homes, is what catches Frank's attention.

The saucer tracks the Grown Homes thread for the next few years, and finds a most remarkable dwelling: the Heironymous Bosch House in Woodside, California, constructed for none other than Biobot founder Saleem Irawaddy. The home's design is inspired by Bosch's famous painting *The Garden of Earthly Delights*, a work that both Saleem and Uli admire.

Instead of being a Giga Gourd, the Bosch House is constructed from the seed cases of several giganticized mutants akin to the Chinese lantern plant, or winter cherry. Five seed cases are combined. A four-sided silvery translucent seed case is the bedroom, an eight-sided clear transparent one is the dining area and living room, a five-sided shiny bright-red one is the kitchen and bath area, and Saleem and his wife Leela's offices

FIGURE 39: The Bosch House

are in a six-sided dun-brown seed case. The four pods are securely mounted on concrete foundations, with springy polyglass tubes connecting them. A herd of snow-white unicorns roams the grounds, as well as a flock of demonic rhamphorhynchii. There are scaled-up songbirds and blackberries as well.

What truly sets the Heironymous Bosch House apart from the primitive Casas Gordas is that, thanks to some heroic gene-tweaking, the component seed pods come with plumbing and wiring already grown in. That is, the actual veins of the pods are just the right size and shape to use for piping, and thanks to a splice with the knifeplant genes, the seed cases have a nice filigree of metal wires, well-insulated by ribs of pith.

Beneath the Sea, upon the Moon?

The next step is for people to live in giant plants that are still alive. Like insects do! Grown Homes produces a wonderful line of summer cottages that are based on the prickly-fruited springtime vines known as wild cucumbers. The houses are called Green Balls.

Frank and the aliens watch as a Santa Cruz hotelier named Kip Robinson plants a row of five Green Ball seeds in the lawn of his motel, which is at the edge of a cliff that drops thirty feet down to the waters of the harbor.

A month later, five big Green Ball pods have appeared, complete with transparent hull sections that act as windows, great soft pith beds, and conical absorbent toilet holes. All summer, families come to spend time in the Green Balls. The air within them is fresh and oxygenated; the self-renewing inner surfaces are plump and full of sap.

Towards the end of the summer, a vacationing couple from Virginia do some wild partying in their Green Ball, and in the course of things the big, ripe pod comes loose at 3:00 A.M.—and rolls off the cliff to plop into the ocean. The nosy saucer follows the tumbling Green Ball to watch. Owing to the softness of the inner walls—and perhaps to the limpness of the wasted Virginians—the couple aren't injured. In fact, they barely notice the mishap. The woman, Diana, pulls the door flap closed before much water can slop in. She thinks it's raining. She and her husband Henry feel dizzy and lie down to sleep. The Green Ball floats in the bay until dawn breaks and a fishing vessel spots it, half a mile offshore.

Far from wanting to sue Kip Robinson for renting them the Green Ball, Henry and Diana are enchanted with their experience. And Kip gets the idea of making houses that *belong* in the water. He loves the ocean, and this seems like a perfect idea. Kip raises some more money, cuts a deal with Grown Homes, and pretty soon a bioengineered "Kip Kelp" seaweed has been developed to provide underwater houses called Sea Homes.

Just like any kelp, the Kip Kelp has a bunch of roots called a *holdfast* that attaches to rocks on the seafloor. Kip Kelp has a long, tubular stalk that runs from the seafloor and abruptly bulges out into a beet-shaped bladder that is naturally filled with air by the respiration of the plant. In normal kelp, the bladder might be a foot across; the Kip Kelp bladder is thirty feet in diameter.

FIGURE 40: Kip Kelp Sea Homes

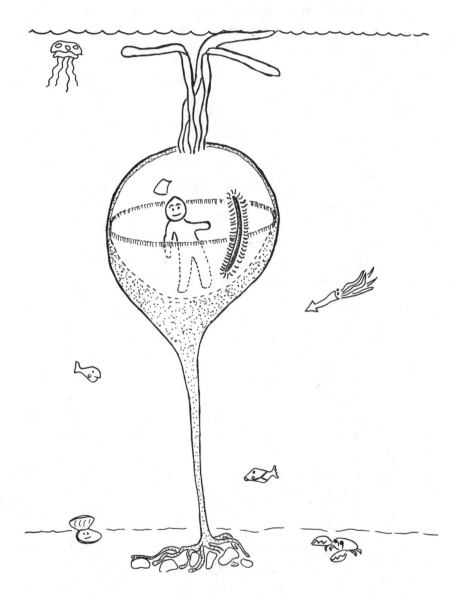

Frank and the aliens watch as Kip Robinson himself tries spending a night in the prototype Kip Kelp Sea Home. Getting inside the bladder is a struggle in itself, as it's floating twenty feet beneath the surface. The wetware engineers have designed a system of interlocking flaps a bit like an airlock, but Kip finds getting through the lock to be a slimy, intimate process comparable perhaps to being born. Inside the opaque pod everything is slippery and wet. Just to prove that it can be done, Kip toughs out the night, but in the morning his opinion of his first Kip Kelp Sea Home is starkly concise.

"It sucks."

And then he gets a mild case of the bends and has to go into a recompression chamber for forty-eight hours.

Frank and the saucer check in a year later, and by then all the bugs have been ironed out. The upper half of the floating Sea Home bladders are transparent. The door has become an easily-navigable sphincter. And the lower part of the bladder is now pith-covered and dry. The difficulty of swimming down to the pod and having to decompress on the way up is solved by having the Kip Kelp's stalk length fluctuate on a twenty-four-hour cycle: mornings and evenings the floating pod is right at the surface, while nights and middays it's down low.

Of course, there are still some inconveniences, but a number of people choose to live in Sea Homes, enough of them that submarine zoning becomes quite the local issue.

Frank is tiring now, but Herman wants to show him more, much more. Trying to turn off the flow of information, Frank focuses on his porringer of gruel, upon his mug of water. To no avail. Quick glimpses of Grown Homes in space float by, and there are views of giant, green domes growing upon the surface of the moon. Frank wonders where they're getting their water, but he doesn't bother inquiring.

"I've had enough," says Frank. "This is more than I'm going to be able to remember."

"But I still want to show you what people will do to themselves," says Herman.

Frank feels a flicker of curiosity.

"All right, but make it fast," he says, sighing.

"Just a preview," says Herman

MORPHED HUMANS

I'm so tired. Too much. Morphs. Don't know how they do it. Four boobs, kangaroo tail, ape, flower hand, canyon Batman, polar people, double man, dickhead, tough pod thing—is that a person? Enough enough enough.

What Frank sees now is a rapid montage of images, some of them seemingly drawn from very far into the future. There is no time to get into the individual stories about the people he sees, or to find out the steps they used to alter their bodies. It's all that the weary Frank can do to assimilate and remember nine of them.

• A woman with four breasts. She wears a dress with a complicated double décolletage. She's in a nightclub with a short man who has little devil horns growing out of his bald head.

• A man with a big kangaroo tail. He jumps backwards, lands on his tail, uses it to bounce himself forward. He is in an Earth city, but the buildings seem impossibly lacy and tall.

• A woman with long arms like a gibbon. She swings through trees, musically hooting. On closer inspection, Frank can make out that each of the trees is a home, hollowed out like a fairy-tale tree house. Nobody but this one woman has been changed into a gibbon. She peers into the lit windows; her song sounds lonely and disturbed.

• A newborn child with an augmented left hand. Each finger branches into five smaller fingers that rebranch, down through four levels for an end result, says Herman, of 625 fingertips. The minute pink fingers stir like the tentacles of a sea anemone. The father is happy and excited, the mother is upset and in tears.

• A man who has a tiny, shriveled body and huge, leathery wings. He hang-glides on thermals rising from the Grand Canyon. Silvery lens-shaped aircraft float in the background.

• A couple who are covered with pelts of thick, white hair. They are sitting on an ice shelf, eating a large, raw fish. Their feet are webbed and their arms end in red claws. Skirling sea-birds wheel above them, hoping for scraps.

• A man who is two times normal size in every dimension.

FIGURE 41: Nine Morphs

His limbs are thick and squat to bear the eightfold extra weight. He has a ruff of stegosaurus spikes down his back to radiate off his extra heat. His great cow eyes glitter with intelligence. He is eating a giant apple.

• A truck driver whose head is a wimpled roll shape, like a foreskin, thick meat whorled around sushi-style, with a glistening eye in the center. On second glance, that's not a truck he's driving, but some kind of spaceship.

• A person who is a tough-skinned spindle pod floating in outer space. Algae within this person's flesh provides oxygen. He or she moves about via a finely directed rocket jet of some kind. At the forward end of the spindle is a single thickly coated eye.

The parade of freakish morphs make Frank so queasy that he faints. This is enough to finally convince Herman to let him go back home.

As Frank comes to, he's looking down at a house in Los Perros, at Rudy Rucker's house, looking in on me as if the walls of my house were transparent. The aliens show him a few things about me—but Frank won't say what—and then they zoom over the Santa Cruz Mountains and he's above his dear little house in San Lorenzo. There's a downwards rush, and at last Frank's back in his chair with his desk and his three televisions.

EIGHT

The *Mondo* Party

I ASK FRANK

I finished writing up the biotechnology notes on Tuesday, June 28, 1994. On the morning of Wednesday the twenty-ninth Frank called me again. I felt a little guilty about the sour note on which our last encounter had ended, so I decided to make a friendly gesture—which would prove to be a big mistake.

"How would you and Mary like to come to a big party with Audrey and me tomorrow night?" I asked him. "*Mondo 2000* is throwing a bash up in Tilden Park near Berkeley. They've rented a space called the Brazilian Room."

"What's *Mondo 2000?*"

"Man, you're *way* up in the mountains, aren't you? *Mondo* is this cutting-edge, totally hip magazine published by some freaks in Berkeley. Instead of being about acid it's about computers. It's cyberpunkadelic."

"*Computers?* What's hip about them?" demanded Frank, sounding querulous. It was like he had no understanding at all of the "Notes on the Future of Communication" material—even though he'd been there with the aliens to see it all coming true.

"Come on, Frank. Think about Larky's Brain Concert. Think about sluggies and radiotelepathy. That stuff builds on our whole computer-based technology. That's where computers are going to lead us."

"I wouldn't call any of those future things computers," said Frank stubbornly. "Computers are brittle beige boxes. They suck. I hate them."

"Computers are a little boring now, Frank, but they're going to evolve. As you of all people should very well know. *Mondo 2000* was the first magazine to see it coming."

"Do you work for them or something?"

"I helped edit an anthology of their best articles called *Mondo 2000: A User's Guide.*[10] But I don't write much for them anymore. If I mail them something, they always lose the manuscript the first time, so I mail it again, and then two months later I get a frantic phone call from some editor with their voice all weird and they want me to fax an immediate rewrite because they hate what I wrote and the deadline's tomorrow, and I do that and they lose my fax, and I fax it again, and then four more months go by and the new issue comes out, but they don't mail me a copy because they've lost my address. And they won't even *discuss* the subject of paying me. On the upside, *Mondo*'s gotten me a lot of good publicity, and thanks to my agent, I *did* get paid for the *User's Guide*. In any case, it's a happening scene, and the Mondoids definitely know how to throw a good party."

"When is it again?"

"Tomorrow; Thursday night. Audrey and I are going to drive up there around eight."

"Let me ask Mary."

In the background I could hear Frank and Mary talking for what seemed like a long time. While I was waiting, Audrey wandered by and asked me what I was doing.

"I'm on the phone with Frank Shook. I invited him to come to the *Mondo* party with us, and he's asking his wife."

"Oh, great," said Audrey. "We're going somewhere with a person like that? Really, Rudy, that is so lacking. But I'm sure the *Mondo* people will love him."

---

[10]Rudy Rucker, R. U. Sirius and Queen Mu, eds., *Mondo 2000: A User's Guide to the New Edge*, New York: HarperCollins, 1992.

Just then Frank was back on the line, all enthusiastic.

"Sounds good, Rudy. We haven't been to a party in a long time. What if we drive up to Los Perros and catch a ride to Berkeley with you?"

"Yeah, we could give you a ride. Where should we meet?"

"Not here!" hissed Audrey.

"Our house is a little tricky to find," I said smoothly. "We could meet at the Los Perros Coffee Roasting Café. Right in front there at eight o'clock. Try not to be late!"

"Can do, chief."

I hung up and chuckled. "That was pretty slick how I put him off, eh, Audrey?"

"I hope he's not chuckling at *you* right now," she said. "It would be terrible for him to know where we live."

So around eight-thirty on Thursday, June 30, 1994, we set off for Berkeley from Los Perros in my Acura Legend, Audrey and I in front, Frank and Mary in back. Although I didn't tell Frank, I had the manuscript for *Saucer Wisdom* in my knapsack with my sweater and my extra glasses. I still wasn't planning to show it to him, but I wanted to have it handy in case something he said would make me want to make a quick correction.

"Have you lived here long?" asked Audrey as we pulled onto the freeway.

"I grew up in Santa Cruz," said Mary. "And Frank's been here for . . . is it twenty years, Frank?"

"Twenty-two," said Frank. "I grew up in Wisconsin and when I got out of high school in '68 I joined the navy. I signed up for a four-year tour in the North Atlantic so I wouldn't have to go to 'Nam. I worked in the radio shack. Just call me 'Sparks.' I mustered out in Newport News in '72 and headed straight for San Francisco. 'Be sure to wear some flowers in your hair.' "

"The peak of the sixties," I put in.

"Frank and I were in an urban commune together," said Mary. "In the lower Haight. That's how we met. We were raising pygmy goats in the backyard and making yogurt. But there was always hair in the yogurt, and nobody would buy it, and we didn't want to eat it, either. Our leader was named Kenny Natur. He was really ugly but he had this great smile. Thick lips and a beard that made him look like a yak. He was always trying to get us to do Tantric orgies with him. Frank was our media director for about fifteen minutes.''

"I had a reel-to-reel tape deck,'' said Frank. "And seventeen miles of music. But then it all got ripped off and I blamed Kenny and I did something stupid. I had to go away for awhile, and when I came back I got a job working for a TV-repair shop in the Mission. And a few months later I convinced Mary to move in with me. We had a room the size of a closet. I think maybe it *was* a closet. There weren't any windows.''

"We weren't really ready to be together yet,'' said Mary. "So I moved down to Santa Cruz and got a job at a crafts shop. I was dating a bunch of other guys for quite a few years.''

"I missed her so much I eventually moved to Santa Cruz too,'' said Frank. "At first I worked as a groundskeeper at the university. Sweeping up after the eucalypti. And then I was working with Spun at the garden shop. After Spun and me got busted, I got a job as a tech in the language lab back at UC. And then came Western Appliance. Mary kept refusing to live with me again, but finally I got her to marry me in uh—''

"Eighty-nine,'' said Mary. "It was the same year you and Peggy Sung first saw the saucers. Remember? You said you'd had a vision of us two living by San Lorenzo. And you were right.''

"That must be exciting to see UFOs,'' said Audrey. "I'd like to see one.''

"Yes, Frank sees saucers all the time,'' said Mary. "And they even take him for rides. But not me. I'm too Earth-grounded; I'm a Taurus. Do you think Audrey could see a saucer, Frank?''

"I don't know her well enough to say,'' said Frank.

"I've been wanting to ask you, Frank,'' said Audrey. "Are the aliens anything like jellyfish?''

"None of them that I've seen so far are like jellyfish,'' said Frank. "But they come from all over, so some of them might be. I've seen some that looked like scarab beetles, and some like starfish with big, red water balloons on them. You didn't say what's your astrological sign.''

"Is this a disco?'' said Audrey, laughing. "I'm Aquarius.''

"Aquarius is an air sign,'' said Mary. "That's good. Do you ever have dreams about flying, Audrey?''

"Once in awhile," said Audrey. "It's my favorite thing. The best one I ever had, I was flying in the snow. It was so wonderful. It wasn't cold, it was warm and balmy and I was in my nightgown flying through falling snow. Beautiful flakes flying right towards my eyes. I'd like to paint that."

"Audrey's a great artist," I said. "She's having a show of her jellyfish paintings starting at the Los Perros Coffee Roasting on Saturday, as a matter of fact. We're going to hang the pictures tomorrow night."

But Frank and Mary were a little too cloddy, or too redneck, to follow up on that. "Frank has a lot of flying dreams," said Mary, almost as if Audrey and I hadn't said anything.

Audrey and I exchanged a look that spoke volumes. But Audrey politely did her part to keep the conversation rolling. "Are you asleep when you see your UFOs, Frank?"

"No, no," said Frank. "I'm always wide awake. Hasn't Rudy told you about how I do it?"

"I thought you might want to keep it confidential," I said.

"We're planning to publish this, aren't we?" said Frank. "I've been scared to talk all these years, but now I'm talking and nothing bad is happening, so, hey, I'm ready to tell everyone I see. I don't see why you're such a tight-ass about showing people what you've already written. Will the *Mondo* editor be there tonight? Maybe you should let them publish one of your chapters in the magazine."

"Well—like I said earlier, Frank, I'm not into writing for *Mondo* anymore. But let's back up for a minute. I was there with you the other day when the saucer came for you. If it's real, how come I didn't see it? Please don't tell me this has something to do with astrology."

"Astrology is a scientifically tested method for classifying personalities," said Mary.

"Maybe so, but it has nothing to do with UFOs," said Frank. "Like I'm always telling you, the aliens come from completely different parts of our galaxy or even from other galaxies entirely. The zodiac, lest we forget, is based on the constellations visible from planet Earth. No, don't worry, Rudy, I'm not going to turn New Age on you. The reason you didn't see the saucer is that it was only in the room for about a microsecond of your time. It's like—don't blink or you'll miss it. And you blinked. Most people do."

"So how do *you* manage to see it?" asked Audrey.

"I see it because the aliens are interested in me—on account of me being such a brilliant, stand-up guy—and they bend my timeline off into paratime with them. UFOs almost never let themselves be visible in regular time. Usually the only way you see one is that they bring you into *their* time dimension. Everything around you stops except for the saucer. It could happen to us right now."

There was a silence then, all of us kind of waiting for the aliens to arrive, but nothing happened. For some reason I thought of the Magritte painting of a galloping horse on the roof of a car speeding down a highway.

"Or we could fall asleep from boredom," said Audrey. "Just kidding. Did you say you used to work in a garden shop, Frank?" Frank did know a lot about gardening, and he gave us good advice on how to take care of the yucca plants on our deck. From yuccas we moved on to deserts and beaches, with a satisfyingly detailed four-way discussion of the beaches of the world that we'd visited. Then Mary and Audrey got onto the topic of sharks, whom Audrey hated just as much as she loved jellyfish, and then we were in Berkeley.

THE PARTY

When we got to Tilden Park it was just getting dark. The Brazilian Room was deep in the park on Wildcat Canyon Road. It was a long, low wooden hall surrounded by a big patio, perched on the brow of a meadow rolling down to some woods. It was a Casbah party, and a lot of people were wearing odd clothes.

Wes, the titular editor of *Mondo*, was dressed as an Arab woman named Amara. The Casbah theme had been Wes's idea. He felt that we should think of the Web as an arabesque labyrinth. The *Mondo* editors were big on understanding technology as metaphors—instead of actually learning anything hard. *Mondo* owner and chief editor Queen Mu was holed up in the kitchen, inaccessible behind starry eyes and rictuslike smile, her voice breathy and brittle, *stay away*. The *Mondo* art director, Bart, a saturnine man with a shaved head, was nicely turned out in a tuxedo and an authentic-looking maroon fez. *Mondo* cofounder R. U. Sirius was slouching around on the patio with his lovably goofy grin, eternally fingering his long hair, simultaneously alert and bemused.

"Hi, R. U."

"Rudy! What's happening? Hi, Audrey. Glad you guys could make it."

"This place is magical," said Audrey. "I never knew it was up here."

"Find your way through the labyrinth," said R. U. "I think I was at an acid test here a really long time ago. Or maybe it was a wedding."

"This is my friend Frank Shook and his wife Mary," I said. "And this is R. U. Sirius. He's *Mondo*'s—I think the last issue's masthead lists you as 'Icon At Large'?"

"Very iconic, very large," said R. U. "Are you a mathematician, Frank?"

"What an insult," laughed Mary. "Can you imagine *looking like a mathematician*? Whoah."

"I can imagine it all too well," said Audrey, giggling. "You should see some of Rudy's friends."

"I'm a saucer abductee," said Frank. "Rudy and I are working on a book."

"Wow. That's quite a departure. Do you guys have a publisher?"

"Not yet," I said.

"Rudy doesn't want to show anyone the manuscript," said Frank. "Not even me. For all I know he's not even writing anything."

"That's the best way to work," said R. U., laughing, and then someone else came up and started talking with him.

"Let's check out the rest of the party," I suggested, and the four of us merged into the throng. There was a good spread of Mideastern food, lots of beer and wine, a fair amount of pot out on the fringes of the patio, and who-knew-what odd chemicals in some of the people's bloodstreams. For entertainment, there was a celestial hippie-dippie storyteller, a cursing-poetry performance, and bizarre, instrumentless electronic live music.

Among the guests was an astronomer from U.C. Berkeley; it turned out he'd read one of my science books. I introduced Frank to him, describing Frank as a friend of mine who was interested in extraterrestrial intelligence.

"Do you think aliens could travel as cosmic rays?" asked Frank, getting right down to it. Some wine sloshed out of the plastic cup he was holding.

"That is a genial notion," said the astronomer. He was a professor on a sabbatical visit to Berkeley from his native Rome. "What do you have in mind, exactly?"

"I think—well, actually I *know*, but let's not get into that." Frank chuckled mirthlessly. "I think beings all over the universe code themselves up, both mind and body, and send the coded patterns across space as high-energy electromagnetic signals. Cosmic rays." Frank held out his arm and wiggled his fingers, and then drew his fluttering hand down an imaginary line towards his head.

This being a *Mondo* party, the professor was primed for weirdness, and he took this happily in stride. "Like Puck who slides down a moonbeam. A most efficient method for interstellar travel. One caveat: most so-called cosmic rays are bits of matter, something like an iron nucleus or perhaps a very energetic proton. A particle like that is not carrying much information. Really, just atomic weight and the energy-momentum four-vector, a few bytes, nothing. But some cosmic rays are energetic gamma rays. These are the signals your extraterrestrials can be modulating with information. Even better are the cosmic GRBs, which stands for *gamma ray bursts*. Yes, the GRBs are the best candidates for your alien moonbeams. GRBs last about a second and contain, oh, many trillions of gamma ray photons with a widely varied and complex structure."

"Nice try," said Frank, condescendingly. "But the correct answer is that the aliens' electromagnetic signals have a higher-dimensional component. What we see is just the tip of the iceberg. You should show him our chapters, Rudy." And he went to get more wine.

"Is your friend a UFO contactee?" the astronomer asked me after we'd chatted a bit more.

"He says so," I answered. "I'm working on a book about his experiences. So much for my academic credibility."

"Oh, well. Life is to make entertainment," said the Italian astronomer. "Your friend knows this."

Looking across the patio, we could see Frank deep in conversation with a very impressive dominatrix-dressed woman. He was throwing back his head and laughing, while making rapid circling gestures in the air with his hands.

"Perhaps he offers her a ride in the flying saucer," said the astronomer. "She is very sexy."

"I'm not sure she's really a woman," I said. "She's from San Francisco."

"America is the great labyrinth," said the astronomer.

The evening wore on. The air was filled with a combination of licentiousness, California weirdness, and business chatter. There were lots of young gen-Xers, tons of old punks and hippies, and a good sprinkling of New Wave media artists, writers, and scammers from the ages in-between. Everyone had the sense that somehow, some way, there was money to be made off the rising tidal wave of electronic information, but nobody was sure how. Above and beyond all that was the joy of being at a big, weird party, a feeling of being close to the center of hip.

I noticed Frank talking for a long time with a British-accented pup who styled himself as the publisher of a new *Mondo*-clone magazine out of London. I'd met him before, and had pegged him as a rip-off artist and a wanker, unlikely to ever even put out one issue, let alone pay anyone. His name was Nigel. Frank briefly drew me into their conversation.

Nigel was playing the multimedia mogul, and Frank was falling for it, even though Nigel was incredibly drunk. Frank wanted to sell Nigel serial publication rights for my chapters on the spot. I demurred and wriggled away. Behind me, I heard Frank saying, "Rudy's trying to hold out on me, but he'll let me have the manuscript pretty soon. Sooner than he thinks!"

This last phrase was shouted after me. It made me nervous. When I was sure I was out of Frank's sight, I checked that my manuscript was still in my knapsack, and then, just to be on the safe side, I locked my knapsack into the trunk of my car.

About an hour later, an ancient VW Beetle veered off the road, puttered across the meadow and pulled up at the edge of the patio. And

who should emerge but—Spun and Guster. Spun was carrying his conga
drum, and Guster had a big cardboard box.

"Oh, no," I said to Audrey.

"You know them?" she said. "They look like homeless stoners."

"They're friends of Frank's. I never should have invited him." I
looked around, not seeing Frank anywhere. "Have you seen him lately?"

"The last I saw of him, he was trying to sell something to someone,"
laughed Audrey. "Something big that he had in his pocket. And then he
was over by the bushes yelling, but I couldn't see who he was talking
to. I've been making a point of telling everyone that you're really good
friends with him."

"William Gibson!" hollered Spun, spotting me. "I brought the
cunga! Budda boom budda bammity bip bip bip zee zow zee zow zee
zow zoooooom!"

" 'Sup, Rudy," said Guster. He and Spun came straight over. Spun's
eyes were completely bloodshot and Guster's beard was wet with red
wine. "Where's the bar at?"

"Inside," I said. "What's in the box?"

"Frank asked me to bring up a bunch of Lotus Lights," said Guster.
"He figured these Berkeley types might like 'em. Keep an eye on our
stuff, would you?" He set the box down on the edge of a planter and
Spun left his conga there too. They started in towards the bar, but got
distracted by the big San Francisco dominatrix. Maybe she *was* a woman.
Maybe it didn't matter.

"Duuude," Spun told her. "I'm gonna play the *cunga*."

"The Lotus Light is nine out of ten gynecologists' choice!" hollered
Guster. "Up, up in the stirrups, and awaaaay!"

"What are Lotus Lights?" asked Audrey. "Are they ray guns?"

"I'm not sure." We looked in the box; it was full of flashlights that
had a kind of plastic cone glued over the lens. Audrey turned one on
and so did I. There was a weighted wheel inside the cone that moved as
you moved the light; the effect was that the color of the light kept changing. A few friends joined us and we were all waving the colored lights
around, laughing.

"*Far out*," said Audrey, not quite seriously. "Cosmic!"

"Ten dollars each!" shouted Guster, puffing back from the bar with
a glass of beer and a glass of wine. "Yes! All you Bezerkely freaks

need the Lotus Light to see God! Test trial is free, but don't be runnin' off with them. Dr. Spun will provide the audio section of our presentation.''

Spun fired up a huge Santa Cruz spliff and began to drum. Guster began juggling three of the Lotus Lights. I would have thought his coordination would be shot, but he juggled as easily as if he were moving in slow-motion. Now Frank and Mary appeared, walking up from the misty meadow. They looked happy and relaxed.

''Hope you don't mind that I told Spun and Guster to come, Rudy,'' said Frank, sitting down next to us. ''I like to cut my brahs in on the good times. And this seemed like an excellent opportunity to sell some Lotus Lights.''

''I guess I don't mind. Spun and Guster are pretty funny to have around.''

''You're not worried any more about them spying on you?'' said Frank. As he said this, Guster put an extra double-flip onto one of the bobbing Lotus Lights and Spun beat an extra tattoo on his conga.

''*Are* you spying on me?'' I asked Guster, but he just grinned and kept juggling.

It was getting cold now, so Audrey and I went back inside the Brazilian Room. These days I was trying not to get wasted at parties anymore. I was unable to predict which times I'd be successful, but tonight King Alcohol and Queen Jane were letting me off easy. Tonight I was going to be okay. We ate and talked and danced until it was past midnight. And then Mary appeared.

''Frank rode home with Spun and Guster,'' she said. ''One of the *Mondo* people started bawling them out for selling Lotus Lights. Some crypto-yuppie. I still need a ride from you. I'm supposed to pick up our car. Do you want to go soon?''

''Why isn't Frank riding with us?'' I asked.

''I don't know,'' said Mary. ''I guess he wanted to hang with Spun and Guster. Or he didn't want to bother you.''

''Well, okay.''

BURGLARIZED

The ride home was uneventful; Mary and Audrey slept most of the way. We dropped Mary off at her car in front of the Los Perros Coffee Roasting and drove to our house. I got my knapsack with the *Saucer Wisdom* manuscript out of the trunk and we went inside.

Right away Audrey noticed something was wrong. "We didn't leave all those lights on, Rudy. Someone's been in here! Look, the rug's all crooked! My jewelry! Go upstairs to our bedroom and see. No, don't! They might still be in there!"

We fell silent and listened. The house was absolutely still.

"There's nobody in here," I said. "I'll check the bedroom."

"Should we call the police?"

"Let me look around first." For protection, I took the hammer out of the kitchen toolbox. Then I went upstairs to our bedroom. It was dark. I had a fleeting spasm of fear that when I turned on the light I'd see an alien in there, maybe one of those repellent starfish that Frank had described. Clutching the hammer, I thumbed the light switch. The lit-up room was blandly unharmed, all the furniture just as it always was, like "Hello, I'm your and Audrey's bed, you sleep and make love on me," "Hello, I'm the chair where you pile your dirty clothes," "Hello, I'm Audrey's dresser, I've been around for years and you don't really know what's inside me," "Hello, I'm—"

"Your computer!" Audrey cried from downstairs. "My paintings are okay, but your computer's gone!"

Sure enough, my computer was missing from my little office off the kitchen. Someone had taken the whole system: printer, monitor, keyboard, mouse, and CPU. Even the power cords were gone.

"That son of a bitch," I said. "Frank Shook did this. That's why he left the party early."

"Call the police."

"If nothing else is missing maybe I should handle this myself," I snarled. "I'll drive down to his house tomorrow and—"

"Rudy, don't try and make this a personal thing. You could get killed! If someone breaks into your house you call the police. That's what they're for."

So we called the Los Perros police and a squad car came right over. There was a young woman officer and a small, older officer with a mustache. I didn't immediately tell them that I thought I knew who'd done it. I just told them that there'd been a break-in and my computer had been stolen. I tagged along as they looked all over the house for broken locks or windows—but there weren't any.

"You're sure the front door was secured?" asked the man with the mustache.

"Yes," I said. "At least it was locked when I came home."

The woman officer hunkered down and examined the lock, then stood back up. "This particular brand isn't very hard to pick. Of course, if we're dealing with a lock-picker, we're dealing with a professional burglar. But your computer is the only thing missing. Not all that valuable a thing, in and of itself, a used computer." She looked at me levelly.

"Well, I think maybe the thieves wanted the information inside my computer," I said.

"What kind of information?"

"It's a book I'm working on. I'm a writer as well as a professor. I think maybe the burglar is my coauthor. His name is Frank Shook and he lives in San Lorenzo."

"What's his phone number?"

"He doesn't have a phone, and I can't even tell you his address. It's on a back road. I do know how to get there."

"But he doesn't want to go there alone," Audrey put in. "Can you drive down there with him?"

"Not at this time of night," said the man with the mustache. "And in any case, San Lorenzo is out of our jurisdiction. The county sheriff's office can send someone down there tomorrow. We'll call them in the morning and they'll get in touch with you."

"Another thought," said the woman cop. "Do you keep an extra key to your house outside?"

"Yes," I said. "On a shelf in the garbage shed down by the street."

"Let's go see," said the woman.

Sure enough, the extra key was gone.

"Had Frank Shook ever seen you use that key?" asked the woman.

"No," I said. "He's never been to the house at all. At least I don't think so." It occurred to me that it would have been possible for the aliens to show Frank my key. An unsettling concept.

"I know!" said Audrey. "The first time Rudy visited Frank, Rudy locked himself out and had to get the key. If Frank or his wife followed Rudy home that time, they would have seen him get it." Equally unsettling.

The cops asked some more questions, and then they left. Audrey and I had trouble going to sleep; we felt violated and unsafe. I even wedged a chair against the front door; the lock no longer offered any protection.

I slept badly that night, and in the middle of the night I had a very frightening dream.

I'm in a rutted clearing in the woods. There are some creatures high up in the air above me, they're in a fantastical spindly-legged walking machine; the machine is like a rickety bulldozer, it's pushing down a tree at the edge of the clearing, it's rocking and roaring with the effort— uhhnnnm, uhhnnnm, uhhnnnm. I'm protesting, trying to protect the tree. The creatures in the high machine shine a laser down into my mouth, an intense, precisely vibrating beam. The laser is etching chip-designs into my gums, fast and painful, flickering this way and that—zzzzt zzzzt zzzzt—now it's etching onto my teeth as well, changing me for good. If I don't get away they'll do all of my teeth and I'll have nothing left— but I can't move. This is happening too incredibly fast. . . .

I woke myself up. Oh, my teeth. I was terrified. For the first time I could understand the deep, dark helpless paranoia of the poor nuts who think that the aliens have secretly taken over, think that the aliens have made a deal with the government and we are just their cattle, their lab rats, controlled and dictated to by our implants. I went in the kitchen and drank some water, ran my tongue and my fingers across my teeth and gums, paced back and forth, trying to shake off the dream.

And then I started thinking that maybe it hadn't been a dream at all. Maybe those people I'd always made fun of were right—maybe the aliens *do* come and get people while they're asleep. And if there was anyone the aliens would want to get, it was me. I was asking for it, writing a book called *Saucer Wisdom*.

It was completely dark in the house. Thank God, Audrey was here. The only safe and sure thing was the sound of her breathing. It had been a big mistake for me to get involved with ufology. The lure of easy money had led me into evil territory. I decided I was through working with Frank Shook.

I said a prayer, got into bed and calmed myself by listening to Audrey's breathing. It was a good thing not to live alone.

NINE

Missing Time

LOOKING FOR FRANK

But, of course I wasn't really through, or you wouldn't be reading this book.

In the morning we were awakened by two cops knocking at the front door. What a way to start the day. It was a couple of county sheriff's deputies, both in their twenties. One was a talkative, stocky white man named Don, the other a slender, quiet Hispanic man named Luis. They offered to let me show them where our burglary suspect lived. Don drove, but Luis seemed to be in charge.

I threw on some clothes and said good-bye to Audrey. She was doing the final touches on her pictures for the show today, so she was just glad to have me out of the way. "But be careful," she warned me.

I rode in the back of the squad car, and when we got to San Lorenzo I gave the cops the directions on how to get to Frank's.

"Mr. Rucker, we're going to ask you to stay in the car while we talk to the occupants," said Don as we pulled up by Frank's cottage.

"I understand," I said. "But can you lower my window all the way so I can hear?" Luis nodded, and Don pressed a switch to lower my window.

They walked down the path and the Shooks' yellow bulldog came charging out barking. Don shouted "Hello!" until the front door opened and Mary came out.

"Ma'am, can you tie up that dog?" said Luis.

"She doesn't bite," said Mary.

"Any dog'll bite, ma'am."

Mary disappeared with the dog, then came back out. She looked terribly worried. "Has something happened to my husband?"

"We're looking for a Mr. Frank Shook?"

"That's him. He hasn't been here since yesterday. We went to a party and—" Mary spotted me up in the police car. "That man drove us. Rudy Rucker. Rudy, do *you* know where Frank is?"

"No," I called back.

Mary kept shouting questions to me, so then the cops let me get out and stand there with them. According to Mary, Frank hadn't come back home at all the night before, and she still hadn't heard from him. She wasn't sure where Spun and Guster lived these days. It was possible that the three of them were off on a bender. She said Frank hadn't indicated any plans to break into my house before leaving, and she felt there was no reason to think that he was, in fact, the burglar. "It could have been the aliens," she said. "Or maybe it was Peggy Sung."

Luis didn't say anything about Mary's reference to aliens, but he did ask where Peggy Sung lived.

"Her house is right on the highway through Benton," said Mary. "There's a big sign out front that says 'Peggy Sung's Psychic Readings.' You can't miss it."

We said our good-byes to Mary and went back to the patrol car. "I think this is a bullshit case," said Luis. "We got no signs of forced entry, no witnesses, no proof of stolen goods, and now we're hearing about psychics and aliens. You ever hear of *corpus delecti*, Mr. Rucker?"

"What does it mean?"

"It means there's no investigation if there's no evidence of a crime. We got other things to do today."

"We're not going to Peggy Sung's?"

"You want to ask a fortune-teller where's your computer, you do it on your own time. Sorry, but we got a lot on our plate. You can let us know if anything more develops."

That afternoon I had a man over to change the locks on our house. And that night I helped Audrey hang her jellyfish paintings at Los Perros Coffee Roasting. They looked terrific.

Saturday morning, July 2, 1994, Audrey and I sat around the Coffee Roasting for a couple of hours, looking at her pictures on the walls and kind of seeing if anyone noticed them. I couldn't stop thinking about my stolen computer, though, so that afternoon, I drove alone down to Benton and found Peggy Sung's place. The house was a yellow trailer perched on concrete blocks a few dozen feet off the road. Hanging from a rusty pipe in the ground was a sign.

```
PEGGY SUNG PSYCHIC READINGS

PAST * PRESENT * FUTURE

COSMIC SPIRITS KNOW
```

I pulled into the gravel driveway and got out of my car, but then I was scared to knock on the door. I'm not a particularly courageous person. Frank had been so aggressive when we'd seen the Asian woman on the beach. If that was really Peggy Sung, she must be a pretty tough cookie, or why else would Frank have acted like that? I decided to kiss my computer good-bye. It was high time I upgraded anyway. I got back in my car. But right then the door of the trailer opened. It was indeed the same woman I'd seen on the beach.

"Rudy Rucker," she called, walking over. She was dressed in jogging clothes and sneakers again. "I been expecting you."

I got out of my car so that she wouldn't be standing over me. "It's about my computer," I said. "Someone stole it and I thought maybe you could help."

"I have your computer system in my house." She gestured towards the trailer. "You can take it back right now."

"You stole it?! I'm going to tell the police!"

"Your friend Frank Shook very drunk Thursday night. He the one who take your computer system. His no-good friends help him. Cosmic spirits tell him where you keep your extra house key. He only plan to take your *Saucer Wisdom* manuscript, but he not find it. So he take whole computer system."

"Why did he bring it to you?"

"Frank not so smart. He know I can use this kind of machine. So I print out a copy of your *Saucer Wisdom* file for him. I read it before he take it with him. It very nicely written. Good spelling and attractive format. But it very hard for you to find a publisher, I think."

"Why?" I demanded.

"This kind of story is pretty hard for average person to believe. That why cosmic spirits don't care about you writing it. I had worry they might do something to stop you. Yes, I very worry about that before, but cosmic spirits seem to think your activities are harmless joke. Funny stories about the world to come." She regarded me blandly, not unpleasantly. "No problem. Now you take away your computer system. I no like to have."

"Wait," I said. "Aren't you out to get money from me? And weren't you yelling at us on the—" I broke off, totally unsure if the beach incident had really happened.

"Yes, yes, I see you at Natural Bridges Beach," said Peggy. "I go there all the time. I cannot believe Frank throw a rock at me and I have to run away. He such an *asshole*."

"But—but you did talk about money," I said. "Frank said you were going to try and get money from me."

"Not from you," said Peggy. "From him. A long time ago, when I teach Frank how to call the aliens, he promise to give me half the money he earn from them. And then when I hear gossip about you wanting to write book with him, I remind him of his promise. He very angry about this, but he scared maybe I can tell the spirits to stay away. He know he better pay me half the money you give him for the book. But now I have seen the book and I am not excite. Now you take your computer system."

I followed Peggy into her trailer. The front door opened into a tiny reception cubicle, a masonite-walled roomlet with a chair, a reception desk, and a wall-mounted vitrine that held an array of crystals, coins, amulets, and mojo bundles of Chinese herbs. On the desk of the reception cubicle sat a cordless phone—and a crystal ball on a base of three brass dragons.

Peggy pushed open a door in the back wall of the little cubicle. The door led to a cramped, white–shag-carpeted living room with red upholstered furniture. Squeezed into the far end of the little room were three television sets and three video cameras, two of them on tripods, just like at Frank Shook's house.

Moving nimbly, Peggy pulled back a corner of the rug to uncover a three-foot-square trapdoor in the trailer floor. She tugged the plywood door open, and there on a sheet-metal shelf beneath the floor were the components of my computer.

"I was almost looking forward to buying a new one," I said, gesturing at my machine. "But it's certainly nice to get it back." I carried the pieces out to my car. It took three trips. On the last trip Peggy Sung followed me back outside. Evidently, she was ready to see the last of me.

"So where is Frank now?" I asked her.

"He ashamed about what he do to you," said Peggy. "And he scared of police. I think somehow he run away to hide. Don't worry about him. It was nice to meet you, Mr. Rucker."

"Can I talk to you about the aliens?" I asked, not wanting the conversation to be over. "You could give me a different perspective on the things Frank told me. I mean, just to start with, why do you call them 'cosmic spirits' instead of 'saucer aliens'? Is there a difference?"

"No," she said shortly. "But talking about them is work for me. I save it for my customer. One hundred twenty-five dollar per hour consultation fee. If you like, we make appointment."

"But if Frank's giving you half of his cut, it's really in your interest to help with *Saucer Wisdom*," I insisted. "For free."

At this, Peggy Sung laughed in my face. It was clear she thought I'd never get the book published. In her mind I was a pest on a level with my "friend" Frank Shook—a lowly niche indeed. "Good-bye, Mr. Rucker! You have a very nice day!" She turned and went back inside her trailer.

So I took my computer and went on home. By Thursday, July 7, I still hadn't heard any more from Frank. I phoned Peggy Sung and asked if she'd seen him, but she hadn't. When she realized I was not calling to make an appointment for a consultation, our conversation was over. I still couldn't get over how Frank had depicted this calm, practical woman as such a villainess.

Thursday afternoon, I drove down to San Lorenzo to see if anything was happening at Frank's. Audrey had gotten pretty curious about all this, so I brought her along. To my surprise, Frank's house was empty, unoccupied.

"What a dump," said Audrey as we peeked in the windows at the empty rooms. "Are you sure this is the right place?" Someone had completely cleaned the house from top to bottom; it was ready for new tenants.

"Of course I'm sure," I said, feeling frightened and a little unreal. "Let's talk to the neighbors."

We drove back to the Okies' house at the start of Frank's road and I got out of the car. What was their name? Oh yes, the Gandys. The Gandys' rutted mud lot reminded me of the nightmare I'd had about the aliens. Why was I even trying to look for Frank Shook anymore? Better to give up and forget him. Just like at Peggy Sung's, I wanted to get back into my car and slink off—but I was embarrassed to give up so easily in front of Audrey. Right then a woman hollered out of the crummy little house on stilts.

"Can I help you?"

I couldn't make her out behind the window screen, but I spoke up as best I could. "I'm looking for Frank and Mary Shook?" I said. "Did they move out?"

"Are you a friend of theirs?"

"Yes."

"Hold on."

A leathery bleached-blond woman appeared on the shaky steps from the little shack. "Hi, I'm Sharon Gandy."

"I'm Rudy Rucker," I said. "I was working on something with Frank."

"That Frank," said Sharon, shaking her head and making a hissing sound through her thin lips.

"Where is he?"

"Nobody knows. Mary don't want to live out here in the boondocks alone by herself. So she up and left. The landlord's got a new tenant lined up to move in tomorrow, thank God, so Mary's not getting stuck with the July rent. I helped her clean the place up. She had a big yard sale Sunday and Monday, cleaned Tuesday, and moved to Santa Cruz yesterday. We bought one of her TVs. Don't know why they needed three in a place that small. I've got Mary's address written down inside, but she don't have a phone yet. That's the one thing she's happy about.

With Frank gone she finally can get a phone again. He won't stand for
no phone, you know.''

''Yeah.''

''Do you want Mary's address?''

''No—never mind.'' It was really and truly time to forget the whole
thing. ''Thanks a lot, Sharon.''

SOUTH DAKOTA

Life went on, one damned thing after another. My father died, my dog
Arf died, our oldest daughter moved to the East Coast, our son graduated
from college and moved to Oregon, our youngest daughter graduated
from college too.

Audrey did sell enough of her jellyfish paintings to get a jellyfish
aquarium, and that was a fun thing around the house for awhile—though
hardly a replacement for Arf and the kids. Eventually, Audrey got tired
of the jellyfish, and the dry, empty tank ended up in our basement. She
kept painting, though, moving on to hummingbirds, and then to close-
ups of breaking waves.

In the fall of 1994, I started work on a science fiction novel called
Freeware, a sequel to my novels *Software* and *Wetware*. For me, writing
is very much a spontaneous, intuitive process, and although I hadn't
planned to, I found myself drawing on my notes and memories of the
things Frank told me. The colorful piezoplastic, the uvvies, and the aliens
who travel as cosmic rays all found their way into the otherwise wholly
fictional *Freeware*. I finished writing *Freeware* in February 1996.

Around this period I underwent a big change in my spiritual life.
After years of wanting to, I finally got clean and sober. The key step in
my recovery was that I came to believe that God would help me with
my personal difficulties if I asked. In the past, my idea of God had been
a philosophical abstraction: the One, the White Light, the Cosmos, the
Absolute. But now I came to think of God as a personal force that lived
not in my head, but in my heart. And for the first time I realized that
seeking God might have something to do with trying to become a better
person.

Freeware appeared in the bookstores in May 1997, and at the end of that month I got a call from none other than Mary, now living, she said, near Sacramento. She told me that Frank had gotten in touch with her and that he wanted to see me again. She said that he could explain everything.

"Where is he?"

"That's the hard part. He's in the Black Hills of South Dakota, whatever that is. He wants you to come talk to him."

"Talk to him about what? He stole my computer and disappeared for three years. I don't want anything to do with him."

"He thought you still might want to finish *Saucer Wisdom*. He says you owe him because you used so many of his ideas in your new book *Freeware*. He says it's not his fault he's been gone for three years. Supposedly the aliens did it to him. He says he has missing time. Anyway, Peggy gave your computer back to you, so what's the big deal?"

"It's really upsetting to have someone rob your house, Mary. You have no idea. I mean, sure, Frank was drunk, and sure I got the computer back, but even so. And I think he's entirely capable of doing that kind of thing again. I saw him try to kill Peggy Sung once, you know? Are *you* planning to go back to him?"

"Well—actually—in the meantime I've married another man. After Frank had been gone for two years I had our marriage anulled."

"You went to court?" I asked.

"Well, there never was what you'd call a formal legal marriage between Frank and me. A Hindu hippie friend married us up on Four Mile Beach. He gave us a special piece of dried kelp for the marriage certificate, and to annul the marriage I did a little ceremony and burned the piece of kelp. Even though I let people call me Mary Shook all those years, that was never my legal name. I hung onto my maiden name because deep down I knew Frank wasn't forever. My new husband is wonderful. I'm going to have a baby! It's sad for Frank, but my life with him is over. You're right, what you say about him, and you don't know the half of it. Frank is too crazy to live with. I'm glad to be free of him. I didn't realize how trapped I was. But it's not like he's *evil* or something, Rudy. He just makes mistakes. I think you should go see him. He says he wants to apologize. And he loves his printout of the *Saucer Wisdom* manuscript. He promises he doesn't have any problems with it at all."

"Well, I don't know. Maybe I'll call him."

"He doesn't want me to give you his number. He only wants to talk
to you face-to-face. I think he's scared that if you talk to him on the
phone, you two will just argue. He's really intent on patching up his
relationship with you."

"How would I see him face-to-face?"

"Fly to Rapid City, South Dakota. Frank says that's the only airport
up there. If you decide to go, just tell me when, and I'll tell him, and
he'll meet you."

"What's in this for you, Mary?" I remembered the way she'd
brought a tray of food to Frank right before he showed me the first set
of notes, the ones on the future of communication. By now I'd come to
be pretty sure that Mary had slipped those "saucer-written" notes to
him. And the way she'd referred so casually to "Peggy" just now hadn't
been lost on me. I couldn't really trust any of these people.

"I feel sorry for Frank," Mary was saying. "He's so alone now.
You were good for him, Rudy. *Saucer Wisdom* gave him hope. Promise
you'll at least think about going to see him. I'm not going to tell you
my last name, but here's my number; you can call me back. Say hello
to Audrey for me."

I thought about it for a few days.

With *Freeware* finished, I was once again between book projects.
So I dug out the *Saucer Wisdom* manuscript and looked it over; it still
seemed pretty good. I'd always felt kind of bad about the time and
energy I'd frittered away on it. I hate to leave projects unfinished.

But—did I believe Frank Shook? Well, maybe not—but it sure was
fun to try! Certainly Frank's claim that aliens travel as lightlike signals
made a lot of sense, more sense than any other method of space travel
I'd ever heard of. And his visions of piezoplastic had so enthralled me
that I'd made them a centerpiece of *Freeware*.

The ideas Frank had told me were wonderful, even though he himself
seemed like an unintelligent man with a bad personality. This disparity
continued to be a real puzzlement. Frank was like some peasant shaman
filled with miraculous visions. Who could tell where they came from?
If he wanted to say that flying saucers were showing him the future,
why argue. Better just to enjoy what he said.

But was I still angry with Frank about the break-in? Not really. He'd done something stupid, but so had I in the course of my life, many times. In the end, no great harm had been done. If there was anything to still be annoyed about, it was that Frank had disappeared halfway through a book project I'd put so much time into.

I looked at a map of South Dakota and found the Black Hills in the state's southwest corner. It's where Mount Rushmore is. I noticed the famous old Wild West town of Deadwood near there, not to mention the biker meeting place Sturgis, both of them close to the east-west Interstate 90. It would be fun to go somewhere so different.

Though she'd initially been doubtful, Audrey was won over by my growing enthusiasm about going to see Frank. She knew how much it had bothered me to leave *Saucer Wisdom* unfinished. She herself didn't want to go to South Dakota, but as it turned out, there was a week in June when I'd be home alone while Audrey visited her dad in Europe. So what the hey, why not do it. Fly out to Rapid City on Friday, June 13; fly home on Monday, June 16.

I bought a ticket and called Mary; she got in touch with Frank and called me back. Frank would meet me in Rapid City.

To get to Rapid City, I had to change planes in Denver. The Denver airport was new and enormous; my gate was out at the end of a finger in a satellite concourse. The concourse space was so vast that the sounds were hushed and jumbled; the fading voices of the announcements blended with the whirring of the long, sliding rubber sidewalks. It had the feel of an alien spaceport. Anachronistic propeller planes kept flying by like movie images, clearly framed in the huge windows. Beyond the airstrips, endless, flat plains rolled out into the distance.

At the gate where we boarded for Rapid City, everyone was white white white. They referred to our destination as "Rapid," rather than "Rapid City." They spoke in an odd singsong, a musical Scandinavian accent like in the movie *Fargo*. "I'm hopin' the plane gets here before the electrical storm," sang out a farmer in a billed cap; and a sturdy, doughy, lank-haired young woman chorused, "That's what I'm sayin' too." Her *oooo* sound was drawn out and warbling.

I walked across the asphalt runway to get into a low-slung prop plane. There was indeed a thunderhead approaching across the plains. Beneath it was a great patch of slanting rain. I felt myself very much out on the edges of my perceptions. The crackling radio voices from the pilot's console seemed filled with mystery. The next step of my journey began.

Frank was waiting for me at the gate in Rapid City. He'd let his hair and beard grow completely free. He looked like a crazy hermit—well, maybe not quite that bad. He wore jeans, a buttoned-up cotton shirt, and an incongruous tweed jacket. His smile was broad and his eyes were twinkling.

"Rudy!" he cried. "It's great to see you."

"Where have you been, Frank?"

"I've been here in South Dakota for about a year. And before that I've got two years of missing time. Not really missing, it's more that—" He broke off and looked sharply at the people streaming around us. "We'll talk about this in the car. I don't want to blow my cover."

"Cover as what?" I couldn't resist saying. "Have you looked in the mirror lately? Saucer abductions are exactly what people would expect someone like you to talk about."

"Still sarcastic, I see," said Frank. "You're referring to my hair? It makes me feel safer, I guess. I was kind of paranoid for awhile. I hope you're not mad at me anymore. Like I told Mary, I think what you wrote is just fine. The things you say about me don't matter. The main thing is to get the big story across. Come on, the baggage claim is down this way."

"All I brought was this carry-on."

"All right then, let's go on out to my car."

"Where are we going?"

Frank looked evasive. "I thought we'd camp out."

I stopped walking so suddenly that the person behind me bumped into me. "You think I'm going to get in your car and let you drive me out into the woods? Get this straight, Frank: I don't trust you anymore. I think you're dangerous. I came out here to work on a book, not to end up in a shallow grave. Let's have dinner in a restaurant and I'll stay in a motel."

"The place I want us to camp is really terrific. It's right by Devil's Tower."

"What's that?"

"You remember *Close Encounters of the Third Kind*, don't you? Devil's Tower is that awesome volcanic butte where the saucers land at the end. It's only about a hundred miles from here. The northeast corner of Wyoming. There's a nice National Park Service campground there, with lots of people around. I'm not going to hurt you, Rudy. I would never have taken your computer if Guster hadn't gotten me stoned. I can't handle drugs at all. That's why I never, ever, take them. That one night was an exception. I'm not a violent man.''

"What about when you threw the rock at Peggy Sung?''

"I told you before, I wasn't trying to hit her. You've talked to her by now. Did she act like she thought I was trying to kill her?''

"No,'' I admitted. "Not really.''

"All right then. Now come on out to my van. I've got lots of food and all the camping equipment. We'll have a really good time.''

"You don't want to show me where you live, do you?''

"I don't live anywhere right now,'' confessed Frank. "I'm planning to camp all summer.''

"The light dawns,'' I said. "Okay, then, Devil's Tower sounds fine. And, yeah, I remember now what it is; in fact I'd really like to see it. But—I hope this doesn't hurt your feelings—I think I'll rent my own car. Just so I feel more secure.''

"Whatever you like, Rudy. And, dig it, I've got two tents. Give me a hug.'' Frank gave me a warm, friendly hug. He stank pretty badly.

"Maybe I'll buy a sleeping bag of my own,'' I suggested.

"Great,'' said Frank. "We can get it in Spearfish. There's a good restaurant there, too. The Bay Leaf. You can buy me supper and then we can drive the rest of the way to Devil's Tower. You sure you don't want to ride in my car?''

"I'll feel better on my own.''

Frank was driving an old, white Chevy van with a cracked windshield. He had a mattress in the back and a lot of bags of stuff. I rented a compact car and followed him down Interstate 90 to Spearfish. In Spearfish we went into the Wal-Mart and I bought a cheap sleeping bag. And then we drove down the main drag.

The stores in one block: The offices of the *Northern Hills Advertiser* and the *Rapid City Journal*. The Mile High Club, featuring 5¢ Machines, Pabst Blue Ribbon, Old Milwaukee. A dead store with ghosts of torn-off letters saying LANGERS. A Radio Shack. A store advertising *CDs Tapes Books Coffee Bar*, but in a state of *Total Liquidation*, with *25%*

Off Everything. A jewelry store featuring Black Hills Gold. The Spearfish Bootery. The Global Market, an import store with *Goods From Planet Earth.* Sharps Trading Company, a pawnshop with two tires with wheels, a used weed eater, and lots of antlers.

The Bay Leaf was down a side street. We parked and went in. The place was a nicely retrofitted old-time building. Although I was hungry, my stomach was feeling kind of gnarly. In a moment of boredom and blind greed, I'd slipped up and eaten some airplane food. I felt like a rat who's eaten poisoned corn. But the Bay Leaf had some safe-sounding vegetarian dishes, and that's what I ordered. Frank asked for a steak.

"You're not drinking?" Frank said, noticing that I'd ordered ginger ale instead of wine like him.

"I quit about a year ago. It was the only way for me. I wasn't enjoying it anymore."

"That's nice. Have you turned religious?"

"I do have a different view of God than I used to," I said. "In a way, it's like something you told me that Herman said. The thing about the universe being filled with a God vibration. You don't need any special technology to tune in on God. You only need to open your heart and your mind. God's everywhere all the time, ready to help you."

"Yeah, I know," said Frank. "In fact, the last time I went out with the aliens, this was in April—but I better not get ahead of myself. Let me start at the beginning. Are you ready to listen?"

"For sure," I said. "In fact—" I opened my backpack and took out my laptop. "I'll type in what you say."

"Do it," he said. "We need to get this book *done.* UFOs are so popular anymore. And, face it, nobody but Frank Shook knows what they're talking about. Everyone's crazy or lying but me."

"All of the other abductions are fake?" I asked.

"Sure," said Frank. "That's why those feebs never have any kind of story to tell. 'And then aliens probed my rectum.' Fuhgeddaboutit! But let's stop bullshitting around. First of all, I'll tell you how I got here from California."

"Great."

"On the way back from the *Mondo* party, I stopped at your house and I took your computer. The aliens had shown me where you lived and where kept your extra key."

"Yeah," I said flatly. "The same aliens who followed me home in a car the very first time I went to see you."

"Whatever *that's* supposed to mean," said Frank, momentarily very busy with his food. "Anyway, when I went into your house I only wanted to take my *Saucer Wisdom* manuscript, but I couldn't find it. I was wasted, and Spun was acting weird. First, he wanted to find your pot stash and then when he couldn't find it he wanted to piss in your toaster. Like to punish you for hiding your pot and your manuscript. And then he wanted to steal your silver. I got Guster to drag him back outside. Your dog had just been watching us the whole time, but when he saw Guster wrestling Spun, he started barking. We had to get out fast. No way could I figure out how to print out *Saucer Wisdom*, so I just took the whole computer. I knew it was wrong, and I knew Mary would bawl me out. So I got Spun and Guster to leave me off at Peggy Sung's. Peggy let me sleep on the floor, and in the morning she used your computer to print out a copy of your manuscript for me. I've still got it here." Frank fished inside his tweed jacket and pulled out a well-worn sheaf of papers. "I've been taking notes on the blank sides. I've got some wild new stuff for you, by the way. Femtotechnology! Transhumanity!" He handed the notes to me. I laid them down on the table next to my place setting.

Just then our salads arrived. We worked on those for a minute, me quietly peering at the notes Frank had given me. "Where were we?" said Frank presently.

"At Peggy Sung's," I said, looking up.

"Right. I would have brought your computer back to you that morning, but I didn't have my car. And Mary phoned up from the Gandys' house; she said the cops were looking for me. I got worried you'd have me busted. I don't think I mentioned this to you before, but I've already had two felony counts. One was for something I did to Kenny Natur's car when I thought he was after Mary, and the other was for helping Spun grow magic mushrooms in the greenhouse. I never took them myself, you understand, but we were making good money off them."

"What'd you do to Kenny Natur's car?"

"I set it on fire. So even though my borrowing your computer was just a tiny, tiny thing, with that and the car and the 'shrooms, the California three-strikes law could have sent me up for some major time in the pen. So when I heard the cops were looking for me I panicked."

"And then?"

"I got Peggy to let me use her TVs and cameras to attract a saucer. It wasn't Herman and the starfish, it was some creatures like piles of rope. When I started trying to talk to them they began to etch my brain the way they always do, but I showed them *Saucer Wisdom* and it turned them right around. I asked them to drop me off at Mount Rushmore. Why? I'd always wanted to see it, and I knew it was far away from California. The rope-thing I was talking to said that if she dropped me in a different place from where they picked me up, then there would have to be a time difference as well. Some kind of space versus time bookkeeping thing. I said I didn't care. But I didn't realize the time difference would be two years."

"You were in the saucer for two years?"

"No, I was in there for like maybe two days, same as usual. They showed me stuff about the physics of the future. Really startling. Femtotechnology. I wrote it all down on the back of your manuscript. But then when they set me down by Mount Rushmore I found out that it was June 1996. I had two years of missing time."

"And you've been in South Dakota ever since?"

"Well, as soon as I got to a phone I started trying to call Mary. It took awhile to find her. And then it turned out she'd given up on me. I guess she already told you." Frank's voice trailed off. Talking about Mary made him sad.

Our main courses arrived and we ate for awhile. Frank raised his glass of wine, toasting my ginger ale. "Thanks for coming out here, Rudy. It's good of you. I've been scared to call you. And ashamed. But then in May I saw *Freeware*. I saw how you used a lot of my ideas in there. So I figured you must feel a little bit obliged to me after all. And I asked Mary to call you."

"What have you been living on for the last year? How did you buy your car?"

"Mary sent me a little money. My half of what we'd owned. And I worked for six months at one of the local schools here, call it Black Hills High. I was a tech in their IRC—instructional resources center. That gig got me through the winter, and winters are a *biiig* deal here. The Black Hills High IRC had lots of cameras and TVs, but I didn't have my own key, so I was only able to get in touch with the aliens one

more time, which was great, although it led to my getting fired. And
that's where my second batch of notes for you comes from. They're
about the transhuman condition.''

"But you lost the IRC job?''

"It's all the fault of that feeble shithead who cut off his balls and
organized the Heaven's Gate suicide thing this spring. As if a saucer
were a metal machine that would follow in the wake of a comet. Why
not a water-skier behind a sailboat, asshole? Why not an armored knight
on a horse? Or a jet plane or a sled or a bicycle or a surfboard? That's
how much sense it would make to try and use a tin-can spaceship to
travel across the galaxy. And when I tried to tell the people at the IRC
about it—'' Frank's voice had risen and people were staring at us.

"Calm down, Frank.''

"Would you gentlemen like dessert?'' said the waiter.

"We better take off and make sure we get a campsite,'' said Frank,
suddenly collecting himself. "Give my friend the check.''

So we drove westward into the setting prairie sun. All the rolling
fields were green as could be, tender and springlike. Cows were everywhere. We got off Interstate 90 at a town called Sundance and headed
north on a two-lane road towards the Devil's Tower.

DEVIL'S TOWER

Though I hadn't immediately recognized the name when Frank sprang
it on me, I'd wanted to visit Devil's Tower ever since I saw it on a
commemorative stamp when I was about twelve. Back then I had trouble
believing there could be such a thing. Apparently, the way it formed is
that a pimplelike plug of lava bulged up to the surface of the ground
without quite breaking through, the lava hardened, and over the millennia
the steady little Belle Fourche River carved the dirt around the lava away,
leaving a shape a bit like a giant tree stump.

Following Frank's white van towards Devil's Tower, I kept thinking
of *Close Encounters of the Third Kind*, of the scene where the government has barricaded off the great butte in order to keep average people
from seeing the saucer magic, but the hero and heroine plow through
fences and barriers to force their way in. How persistent and attractive

this myth is. The government stands for—what? Your social condition-
ing? Your self-imposed limitations? What *is* it that prevents most of us
from being able to see the UFOs?

Intellectually analyze things as I might, the stops on my emotional
calliope were pulled flat-out. I was Richard Dreyfuss and Melinda Dillon
fearlessly driving onward, on towards the heart of the mystery, filled
with thoughts like, "I'm breaking through, I'm going all the way, nobody's going to stop me from learning the truth!"

And indeed I *was* here for a UFO investigation, and I *was* pushing
past a lot of barriers in order to finish my book about Frank Shook.
Despite the fear of being called foolish, despite the terrifying dreams,
despite the arguments with Frank, despite Frank's treachery and his dis-
appearance—yes, despite all this—here I was, on the homestretch, and
I was really going to write *Saucer Wisdom*.

The Devil's Tower came into view—unmistakable, wonderfully
stark, a frozen icon of upward thrust. There was a historical marker by
the side of the road. Though Frank drove on, I pulled over to read it.
I'd catch up with him at the park gate. I was sure he'd wait there for
me so he wouldn't have to pay.

According to the marker, there's an Indian legend about the Devil's
Tower. A bear was chasing seven sisters. The girls jumped up onto a
big tree stump to escape. The stump grew and grew as the bear clawed
it; the stump grew so much that it propelled the girls up into the sky,
where they became the seven stars of the Pleiades cluster. More facts:
The Tower is about eight hundred feet from base to top. The first person
known to have climbed it was a white rancher, on July 4, 1893. He
fashioned a ladder by wedging sticks into one of the tower's long cracks.
Since then, tens of thousands of rock climbers have climbed the Devil's
Tower. Out of respect for the Native American religion, it's customary
not to climb the Tower in June.

Back in my car, the radio was on. In the Midwest you can either
listen to country music or you can listen to golden oldies. From coast to
coast, there are no other options. The heavy metal band Great White was
playing "Once Bitten Twice Shy," their big hit. It had never sounded
so good to me. I noticed that they had worked in some fairly convincing
Stones riffs. Ironically enough, the very next song on the radio was a
real Rolling Stones song, "Beast of Burden," as if to demonstrate that

Keith Richards's tasty, tart playing always delivered more than mere "Stones riffs."

Frank—my Keith Richards of saucer abductees—was waiting for me at the park entrance. We went in and drove to the Belle Fourche campground down in a river-meadow at the bottom of a slope leading up to the Devil's Tower. We claimed a campsite and sat on the ground, staring up at the Tower.

From this close I could see that the Tower was not so much a stump with grooves as it was a bundle of columns. The columns were hexagonal, sometimes pentagonal, maybe thirty feet across, and the tower itself was—I counted—some sixty columns wide. I thought about that for awhile, wondering how such a structure would form.

"I've got it!" I exclaimed to Frank after a few minutes. "I bet the columns formed as Bénard convection cells in the quiescently cooling magma."

"Come again, Professor?"

"Like when you see mist on the river in the morning, it's divided up into cells? Or if you get some hot miso soup and let it sit for a minute, there's a big honeycomb pattern that forms in the soup particles? Maybe ten of these roughly hexagonal cells? The thing is, Frank, whenever you have a fluid that's cooling off, the hot stuff has to rise and the cool stuff has to come down. And so that it doesn't have a traffic jam, the fluid self-organizes into cells that are vortex rings. Like smoke rings. Each cell's flow moves around and around, like ants crawling up the side of a bagel and down through the central hole. The warm stuff rises up along the outer edges of the ring, and the cool stuff falls down along the central axis—or it can go the other way around. Either way, it's a Bénard convection cell."

Frank didn't look all that interested, but I kept talking. This was a new idea for me, and I was having fun working it out. "I think it must be that for each fluid there's a certain size convection cell that's the best fit for its density and viscosity and so on. You don't get just one giant Bénard convection cell, you get a bunch of smaller ones of the size the fluid likes. Now, the lava that formed the Devil's Tower was slowly cooling off for a really long time, and while it was cooling it self-organized into a pattern that was a bundle of long, thin Bénard convection cells. Vortex tubes, like with the hotter lava moving up along the edges and the cooler stuff moving down the centers. And it froze into that pattern. Which is why it looks the way it does now. Dig?''

"In this pamphlet it says, 'The cooling igneous rock contracted, fracturing into columns,' " said Frank, looking at the National Park Service folder we'd gotten at the gate. "It doesn't say anything about your cells."

"I bet the cells determined where the fractures went," I claimed. "It makes sense, doesn't it?"

"You wanna talk about physics, have I got physics for you," said Frank. "You should read my femtotechnology notes before it finishes getting dark. I'll set up the tents while you do that. We can talk about femtotechnology tomorrow and we'll do transhumanity on Sunday."

So that's pretty much what we did. Frank really did have two tents: a large three-man job that he slept in, and a mesh-roofed one-man mummy tent that he let me use. He offered to let me have the big tent if I liked, but I liked the idea of sleeping with nothing between me and the stars but a layer of mesh.

TEN

Notes on Femtotechnology

There are actually two separate trails around the Devil's Tower: the Red Beds Trail and the Tower Trail. The Red Beds Trail is a low, three-mile path that goes through woods and red-dirt washes, while the Tower Trail is a tight loop right around the base of the great butte itself. Saturday morning Frank showed them to me on a map and explained his plans.

"Let's do this like pilgrims to a shrine," said Frank. "Today we'll circle around the tower at a respectful distance. We'll use the lowly Red Beds Trail. And then tomorrow—if all goes well—we'll approach the central mystery and walk the Tower Trail."

So on Saturday, June 14, 1997, Frank Shook and I slowly walked the Red Beds Trail, frequently pausing to sit and stare up at the great stone shape. And all the while we discussed Frank's notes on femtotechnology.

It had been one year of his time since Frank had taken these notes, so it took him awhile to get going. And it had been three years of my time since I'd last been with Frank, so it took me a few tries to get back my old rhythm of questioning him. I had to jog his memory quite a bit. But in the end, the information flowed as freely and as entertainingly as ever.

MATTER TRANSMUTATION

Woke up at Peggy's, read Saucer Wisdom, it's tremendous! But the cops are coming. I'll get the aliens to take me far away. Mount Rushmore, South Dakota!

Peggy calls a dragon-saucer. Purple reading light. They like Saucer Wisdom. Steffi is a big pile of rope. Femtotechnology built the alla. The ultimate knapsack for the fourth millennium. Rust and Floto take it camping. Steffi fakes their voices for me.

Femtotechnology Unlimited founded by Professor Harry and Surfer Joe. They flip quarks to transmute matter.

The families Robinson in the asteroid. Allas hollow it out. A pond in each end, how nicely they jiggle. Software is your uvvy, hardware is your alla, wetware is your body. An ark from your mouth.

The Rope Aliens

On Friday, July 1, 1994, Frank wakes up on the floor of Peggy Sung's trailer. He's stiff and sore, feeling foggy, and then he notices the stolen computer on the floor next to him. "Oh, *shit!*"

Peggy hears him and comes in from her bedroom. She's wearing jogging clothes as usual. "I ready print out *Saucer Wisdom* file now," she says. "Then you can give back Rudy Rucker his computer system. I no like to have stolen property here."

So they print out the *Saucer Wisdom* manuscript, which is a rough-draft version of most of the material in chapters one through seven. And they sit there in plastic lawn chairs next to the trailer, under the red-woods, drinking coffee and reading.

Frank is relieved to find the book not nearly as insulting as he'd feared. The only things that particularly rankle are where in chapter one it has something about "the termites that live in his gunjy saucer-nut brain," and the reference at the end of chapter five to "Frank's evident mental instability." He really likes the write-ups of his saucer visions.

Peggy doesn't like the *Saucer Wisdom* manuscript, and she's voluble in her criticism. She finds the approach too science-and-technology oriented. She'd prefer to see the saucer beings discussed as cosmic spirits, rather than as traveling aliens. Also, she doesn't like the tone; she feels a book of this nature ought to be more serious. The curse words shouldn't be there, for one thing. And furthermore, she's not so sure it's all right for the book to be mentioning her—but Frank repeats his promise to give her half of whatever money the book earns him.

Around then Spun and Guster stop by to say hello. "You boys very nice to keep an eye on Frank for me," says Peggy. "I glad you bring him here with Rudy Rucker computer system so I know what is going on." She gives them each fifteen dollars and they're off on their merry way.

"You've been paying them to spy on me?" demands Frank.

"Yes," says Peggy Sung. "But it a big waste of money. What Rudy Rucker write—I very disappoint. This book never publish."

It's nearly noon by now, and the phone rings. Peggy picks up the portable handset she has resting by her chair.

"Hello, Mary. Yes, Frank here, he spend night. No way! On my floor. Yes, he have. Cops coming? Oh goddamn hell!" She throws down the phone and runs for the trailer. "Hurry up, Frank! We hide the computer system and you run into woods! Cops coming here!"

Frank snatches up the pages of the manuscript, folds the whole thing in thirds and stuffs it into the back pocket of his jeans. And then Frank and Peggy wrestle the pieces of the computer into the hidey-hole that Peggy has under her rug. Frank lurks in the woods behind Peggy's house for the rest of the afternoon, but the cops don't come. Finally, around dusk he skulks back down.

"You can't stay here no more," Peggy tells Frank.

"Don't worry," says Frank. "I have a plan. The saucer can take me away. Just let me use your rig. I'll have the aliens put me back somewhere else. Somewhere far away."

"You think they can do?" asks Peggy.

"I bet they can," says Frank, turning on the TVs and the cameras. "I'm ready to try."

"Let me handle the controls," says Peggy, taking the handheld camera from Frank. "I no want you screw up my system. You sit here and stare at screen and wait for cosmic spirits. I stand behind you."

"You mean, 'wait for saucer aliens,' " says Frank. "Because that's what they are."

"I hope they take you away and we don't have to argue anymore," says Peggy, twitching the camera this way and that.

As Peggy works the controls, a green-and-yellow knot of spirals appears on the TV screens. The thing looks like—like a dragon. It folds back on itself once, twice, three times, and then somehow it lifts itself out of the central screen.

And now, though Frank hates to admit it, there's a Chinese dragon floating towards him. Its body is made up of bright wire-frame lines just like the saucer visions, but instead of being a nice, clean lens-shape, this baby is a snaky, scaly pop-eyed curvy-clawed triple-happiness golden-luck dragon.

"He coming eat you now Frank," whispers Peggy over his shoulder. "Good-bye."

Tossing its head like a prancing pony, the dragon surges forward, mouth agape. Its bright lines surround Frank and he feels like he's inside a fireworks display with bright tracers shooting every which way around him. And then there's a lurching sensation and he's sitting on an amorphous black bench in a round little room with a transparent floor—just like all the other times the aliens have picked him up. Peggy Sung may have made the craft look like a dragon, but really it's a saucer.

Looking down through the floor, Frank sees Peggy Sung's trailer receding below. They're flying up through the redwoods, up into the sky. The dragon-saucer slopes across the sky and settles into hovering at the familiar location above the San Jose airport.

"Herman?" calls Frank. He has the usual sensation that the aliens are watching him from behind. He whirls around a few times, but no spatial acrobatic he can perform is going to bring the aliens into view. "Let me see you!" Frank calls.

A sudden blast of pain fills Frank's head as a higher-dimensional control beam pokes at his brain. *Zzzzt-zzzzt-zzzzt!*

"Ow!" cries Frank. "Don't do that! I'm not some regular goob who's just here to get a hand job and to be lectured about ecology! Ow ow ow! Stop it! I'm Frank Shook!" He fumbles the folded manuscript out of his hip pocket and holds it up high. "I'm writing a book about you guys to educate the Earthlings! *Saucer Wisdom*! Check it out!"

The etching-ray becomes a shaft of pale purple laser light. The light busily scans over Frank's wad of papers, penetrating the sheets to read every page at once. There's a thoughtful pause and then Frank has a sense of being rotated, as if he's a show car on a dais. Slowly, the aliens come into view.

FIGURE 42: Dragon-Saucer Coming Out of Peggy's TV

They look like a pile of rope—thick, oily, multiple-strand manila rope with a half-dozen loose ends raised up like watchful cobra heads.

"Rope?" says one of the heads, reading Frank's mind. "An all-purpose tool of physical connection and force-transmission!" And then it laughs, a thin, little laugh like an old witch in a German fairy tale. "Hi hi hi hi!"

"Where are you from?" asks Frank.

"We're from what you would call the Andromeda galaxy, Frank Shook," says the high-voiced rope end. "A wetware planet called ♎." The glyph-image is like a ball of twine, like a can of worms. "My name is—oh, call me Steffi."

"You're a woman?" says Frank.

"No," says Steffi. "I'm a live piece of rope from ♎!"

The other pieces of rope flail around in mirth at this zinger, letting out a big chorus of, "Hi hi hi hi!"

"But, yes, you can think of me as a she," continues Steffi. "And, O gnarly-brained Frank Shook, we think it will be amusing to continue your work on the *Saucer Wisdom* notes. So I ask, what aspect of the future do you want to investigate next?"

"I'd like to find out about future physics," says Frank. "There's this word I heard when Herman showed me some people in the early thirty-first century. *Femtotechnology.* I want to find out what that is."

"We can research that," says Steffi. "Although I am already guessing that your *femtotechnology* is what we on ♎ call 'direct matter control.' Let's do like your friend Herman from ♦; let's skim through future UV signals and scan for the target word."

"There's one problem," says Frank. "When I go really far into the future I can't understand the way people speak."

"Oh, then Herman must not have been very good at neurolinguistics," says Steffi. "I can easily improve upon the interface. We ♎ citizens specialize in communication. Like you say, we're ropes that tie things together! Can you give me some mental models of how you like people to speak?"

Frank thinks of himself talking, he thinks of Rudy, he thinks of his wife, he thinks of a surfer friend of his called Wet Willie, he thinks of people on television. Steffi listens to Frank's mental simulations for a few minutes and then she's got them down cold.

"All set?" asks Steffi.

"Sure," says Frank. "Let's do it."

Frank would like to bring up the fact that he doesn't want the dragon-saucer to return him to Peggy Sung's near Benton in the Santa Cruz

FIGURE 43: The Rope Aliens

Mountains on Friday, July 1, 1994. But just for now he's too shy to ask. He's scared that if he pushes for too much at once, Steffi and the other rope-aliens might start etching him again. He figures he can always ask later.

The Alla

They make a big jump into the future, a thousand and eight years—to 3002. And the first mention of *femtotechnology* that they find is in a futuristic UV ad for something called an *alla*.

The alla ad is a completely immersive and interactive virtual reality. For Frank, mentally flying around inside the ad feels just like using the saucer to fly around inside a real-time scene.

Two men are out hiking in the Santa Lucia Mountains above Big Sur. Their names are Rust and Floto. Their clothes are plant-woven cotton, with muted cellular-automata patterns. The sun is gorgeously setting above the foggy, wrinkled sea.

"Okay, Rust," says Floto. "Time to explain. You said you were bringing all our camping supplies, but you don't have a knapsack."

"I'm going to use my Femtotechnology Unlimited alla!" says Rust, drawing something out of his pocket. "The alla can make most anything!"

Thanks to Steffi's seamless subliminal interpretation, it sounds like they're talking 1990s English, a rather bland TV-commercial kind of English at that.

Frank's attention zeroes in on the object in Rust's hand. It's a lumpy little thing, about the size of a golf ball, encased in what looks like green suede leather. Rust makes a coaxing noise with his lips, a kind of *smeep-smeep* sound, and the green "leather" case separates itself from the object and crawls up his arm to perch itself on the back of his neck. The alla-case is, it seems, a piezoplastic sluggie that is also an uvvy. But what of the alla itself?

The business end of the alla is a solid-looking little object made of a tough, shiny metal. This object's shape is that of a tri-bar; that is, it's three short metallic rods that meet at right angles like the corner lines of a room. The little rods are of slightly-unequal lengths, and set into each rod's flat end is a glittering little lens. Attached to the corner of the alla is a contoured handgrip a few inches long.

Unlike all the piezoplastic and biotechnological things Frank's been seeing in the other saucer visions, the alla appears refreshingly hard and angular. It's a piece of masculine fetish technology, if you will, like an old-time fountain pen or pocket camera or ultralight computer.

"What's it do?" asks Floto.

"It makes stuff," says Rust. "Whatever I think of."

"Then how about some dinner," says Floto. "And something soft to sleep on. And some protection against the cold."

"Check this out," says Rust, holding the alla by its grip. Sudden bright lines shoot out of the alla's three blunt metallic rods. The lines are laser-straight, but they're not beams of light, because they go out only a few inches and then they make right-angle turns, growing and turning until there is a bright-edged box of space enclosed by the lines. The region is about the size of a large beer mug.

FIGURE 44: Rust and the Alla

"Dinner," says Rust, holding the alla down near the ground. The space inside the bright box begins to glow. A web of dark lines and surfaces appears in the region, dividing the space up into cellular blocks. The lines warp and buck, subdividing the blocks into smaller cells that break up and divide again and again. There's a whooshing sound and a perceptible breeze as air flows into the alla, feeding in mass for the matter transmutation. The space-patterning grows thick as taffy, opaque and shiny. And then the bright box-lines disappear and a rectangular cup plops down out of the box onto the ground. The cup is shiny-yellow metal, and it's filled with a hot, fragrant broth. The radiotelepathic ad includes smells, and Frank's mouth waters at the rich aroma.

"Mmmm!" exclaims virtual Floto, flaring his nostrils. Rust brandishes the alla and produces a second cup like the first.

"These mugs are an alloy of silver, platinum, and gold," says Rust, tapping one of them. "And the soup contains a complete assortment of human nutrients. I'll make the beds and the covers while our supper cools."

Rust critically eyes the surrounding terrain, then settles on a spot and kneels down. He holds the alla with two of its little rods horizontal and one rod pointing straight down. A bright 3-by-7-foot wire-frame rectangle springs out of the alla, now sketching out a horizontal slab four inches thick. Rust jiggles it like a giant frying pan, then lowers the wire-frame slab so that its bottom rectangle is buried in the dirt and the top rectangle lies evenly with the ground. The rocky dirt inside the bright surface rectangle turns clear, yet lumpy—the box looks like a shallow stream flowing over an uneven bed. A shoal of vortices thickens the material, spinning off smaller vortices like fleas jumping off fleas. Each level of smaller vortex-fleas is duskier than the level before, and then the box-lines wink off and the slab is—

"Like foam rubber!" exclaims Floto, feeling the transformed patch of ground.

Rust quickly fabricates a second rubbery mat next to the first. The sun is fading and the air's getting cold. "Bring the soup over here, Floto," says Rust, and poses grinning with the alla held high over his head like a sword. Like the blade of a sword, the bright-line alla-box is long and thin. Rust makes several passes with the light-sword, and downy, silken blankets fall from the air in its wake—followed by waterproof ponchos.

"Wow," says Floto, as the two sit on their comfortable beds slurping their soup, with soft blankets and ponchos wrapped around them. "You *rule*, Rust. You're *king*!"

"Me and my alla," says Rust, holding the grip so that the bright-line box appears right in front of him. He smiles through the box, and then he digs it into the ground in front of the mats and a tiny campfire appears in a tidy stone pit.

Flickering above the fire is a link to Femtotechnology Unlimited for information on how to order an alla.

Femtotechnology Unlimited

"Let's go back in time and find the start of Femtotechnology Unlimited," says Frank, and Steffi readily agrees. Like the other aliens before her, she enjoys using Frank as a native guide for seeking out stimulating bits of humanity's future history.

Early in the year 3001, they find the Femtotechnology founder, a dumpy guy with thick lips and a slobbering way of talking. He's telling his plans to a tall, skinny assistant who has a little tuft of curly hair. For the moment, Steffi leaves their voices untranslated, and Frank can't even tell where one word ends and the next one begins.

"Make them sound like scientists and surfers this time," suggests Frank. "Like a combination of Rudy Rucker and Wet Willie."

"Fine," says Steffi. "Their names are Harry and Joe."

"We're going to invent femtotechnology now," sloppy Harry is saying. "To make a long story short."

"Say what?" says curly-top Joe.

"I'll explain it again," says Harry. "First of all, here's the official prefixes for small numbers." A radiotelepathically projected chart appears in the minds of Harry, Joe, and the eavesdropping Frank.

NAME	NUMERICAL SYMBOL	PREFIX
THOUSANDTH	0.001	MILLI-
MILLIONTH	0.000001	MICRO-
BILLIONTH	0.000000001	NANO-
TRILLIONTH	0.000000000001	PICO-
QUADRILLIONTH	0.000000000000001	FEMTO-
QUINTILLIONTH	0.000000000000000001	ATTO-
SEXTILLIONTH	0.000000000000000000001	ZEPTO-
SEPTILLIONTH	0.000000000000000000000001	YOCTO-

"Yawn yawn, I've seen that," says Joe.

"Seen but not understood," says Harry. "To grasp the meaning of the word *femtotechnology*, you should first think about the word *nanotechnology*. A nanometer is a billionth of a meter. A big molecule might be ten or twenty nanometers across, maybe a little more. A water molecule is smaller, about a fifth of a nanometer. Nanometers are a natural size-unit for measuring molecules, so when people developed the technology for manipulating molecules they called it *nanotechnology.*"

"Then how come the dooks who work with molecules say they're doing wetware engineering?" asks Joe.

"That's a historical accident," says Harry. "The original nanotechnologists—we're talking about nearly a thousand years ago—thought they were going to be making tiny machines. But that idea turned out to be bogus. Biology has the lock on nanometer-scale fabrication. The word *nanotechnology* died because the first guys to use it had some wrong ideas. It's sort of like the way the alchemists thought substances had philosophical virtues, and then a few centuries later it turned out they'd been trying to do chemistry. The old-time nanotechnologists thought molecules were like machines, and then a few centuries later it turned out they'd been trying to do wetware engineering. Nobody wants to be branded an alchemist or a nanotechnologist because those original groups were wrong in important ways. But my point, Joe, is that wetware engineering is indeed nanotechnology. It's what's going on when you use medi-germs to clean out your arteries. It's what's happening when a diamond-spider spins carbon fibers for construction. It's what happens when a cloth-plant weaves cellular automaton fabric for your shirts."

"Don't hurt yourself, man," says Joe. "You're explaining too hard. Try this one: according to the chart, picotechnology should come before femtotechnology. Why don't we do picotechnology first?"

"There isn't much happening at the picometer-size scale," says Harry. "The next really solid thing below molecules is the nucleus of an atom. And that turns out to be about twenty femtometers. So if we start doing things directly to atomic nuclei we're talking about *femtotechnology.* There isn't going to be any picotechnology because there's nothing interesting that's a picometer in size."

"Very clean," say Joe. "Femtotechnology. I'm down with it, brah. But what are we going to *do* to the atoms?"

"Transmute them, Joe. Dirt into gold. Gold into water. Water into air. Air into chicken soup. Making stuff out of 'thin' air is quite practical, you know. Air has more mass than people realize. A cubic meter of it weighs a kilogram. The air in your bedroom weighs about as much as your body.''

''But how does transmutation work?''

''Transmutation is mostly a matter of changing protons into neutrons and vice versa. An atom's nucleus is a bunch of protons and neutrons. Take oxygen—it's got eight neutrons and eight protons. And hydrogen has one proton. If you could change protons into neutrons, you could stick sixteen hydrogen molecules together, flip half of their protons to neutrons and you'd have a molecule of oxygen. Like that. And, by the way, when you change the nuclei, the electrons take care of themselves.''

''But how do you change a proton into a neutron? Smash it or something?''

''That's the crude, old nuclear physics way. Instead of that, we femtotechnologists are going to use *quark-flipping*. Takes much less energy. What's quark-flipping? A proton is a quark-bag holding two *up* quarks and one *down* quark, while a neutron is a quark-bag with two down quarks and one up quark. To change from one to the other, you just need to go into the bag and flip the one quark.''

''Aren't there other kinds of quarks too?'' asks Joe. ''Besides *up* and *down*?''

''Strange quarks,'' says Harry, smiling wetly. ''We'll get to those later, my man. But first we need to get the femtotechnology matter transmuter working.''

''What do you want to call it?''

''I don't know,'' says Harry. ''Horne O' Planty? Spelled weird to make it a trademark, you understand. Or maybe a Polish Knife? My mother's people are Polish. Or call it an alef? Or maybe a cradle or a loom?''

''How about an *alla*,'' says Joe.

''I like it,'' says Harry after a moment's thought. ''Alla. Fine. Now what's still missing, Joe?''

''I dunno.''

''How to make it work is what's missing.''

''I was gonna say that. How's it gonna work?''

''Incommensurable magnitudes,'' says Harry. ''We'll use three titanium bars of slightly different irrational lengths. The square root of two, the cube root of three, and the fifth root of five will work. Each bar will have a matter-lens that grows quark whiskers, and the whiskers will embody a one-dimensional nonlinear wave pattern that tells them where to turn. The whiskers will split and grow along all the edges of a parallelepiped control volume. Within this box we'll use a chaotic cascade to fuse all the nucleons' quark-bags into a quark-gluon plasma that we're free to flip, shuffle, and regroup. The process will be directed by high-level user-request patterns made via a custom-designed radio-telepathic uvvy that incorporates low-level implementation instructions for a few thousand basic substances. You can help with that part, Joe. And in our commercial release, the control uvvy can act as a carrying case. The alla!''

"Wavy," says Joe. "I'm there, dude."

Life in the Asteroid Belt

Frank and the aliens skim through the ensuing events.

Although it's a shockingly expensive gizmo, the Femtotechnology Unlimited alla is a big hit. Initially, however, its uses are somewhat limited. The first allas are only able to make sheets and blocks of fairly simple substances: The few hundred recipes that Joe managed to program into the control uvvy. But three key improvements follow.

First, the alla control language is provided as shareware to all users, and people quickly add new recipes. Porridge, gasoline, watermelon, tofu, ice, perfume, and—most important of all—piezoplastic.

Second, it becomes possible to program the shape of the alla-box, that is, to design the shape of the region where the transmutation takes place. While the first allas only transmute regions that are either blocks of space or Boolean differences of blocks (like a square cup), Harry and Joe soon find a way to make the quark-whiskers sketch out all sorts of polyhedral and curved-line forms, including spheres and cylinders. Before long, the users can form the alla-box into arbitrary shapes.

Third, it becomes easier to specify different kinds of materials for different parts of the target shape. Although this was already possible in a limited way with the first release of the alla (v., the broth in the gold cup), it now becomes much easier to do. This means it's now simple to make a compound object such as, say, a folding pocket knife. Or a gem-studded silver chair with an embroidered silk cushion. Or even an old-fashioned mechanical watch, although alla-made watches tend not to run—the tension in the springs is never right.

Highly-complex kinds of objects are still out of reach for the alla. For instance, you can't use an alla to make an alla. Or an uvvy. Or a full-grown living organism. Yes, the maps of the wetware engineers can be fed into the alla so as to make custom DNA, and this is in fact a cheaper method of DNA fabrication than the old-fashioned mechanical nanomanipulators. But the ongoing vital processes of a complete, living organism can't yet be captured by an alla blueprint.

Running an alla is expensive. The alla can draw its power from almost any kind of clean energy source, but you need a lot of it. If you need portability, you have to use the quantum dot batteries known as glorks, which also happen to be the international unit of currency. The alla is actually able to transmute, say, a wooden nickel into a glork disk, but it costs a little more than a glork's worth of energy to make the glork, so the basic monetary system is undisturbed.

There is a certain amount of economic dislocation owing to the fact that a kilogram of any kind of substance at all is now worth no more than a kilogram of dirt along with enough power to run the dirt through an alla. Given an alla and an energy supply, diamonds and gold, for instance, are no more expensive than salt and mud. Though, of course, diamonds are cheap already thanks to the biotech diamond trees, whose papayalike fruits have centers filled with moist, sparkling gems.

There is a brief mania for furniture and houses that are made out of (formerly) precious metals such as gold, palladium, rhodium, and yttrium. But the superior comfort of biotech's organic homes can't be denied, any more than the convenience of making furnishings from piezoplastic— especially now that the alla has made the stuff relatively cheap.

The biggest change brought about by the alla is that it finally becomes feasible for people to move out into space in a big way.

The aliens help Frank take a quick uvvy-look through the Net to get a history of space travel thus far. With the coming of the quantum dot energy source in the 2100s it became quite easy to build inexpensive spaceships. The first big off-Earth colony was the Wubbo base at the South Pole of the Moon, which drew on the ice deposits found there, cracking ice into water and air. A similar kind of base was established at the North Pole of Mars where, again, a lot of water ice was to be found beneath the ground. And in the asteroid belt there were some water-rich planetoids as well. The burgeoning of biotech in the latter part of the third millennium made life fairly comfortable in these ice-mining bases.

But it's the coming of femotechnology in the fourth millennium that really opens the door to space. The new technology of matter transmutation obviates the stark, grinding problems of no water and no air. Thanks to the alla, humanity can now move with ease to the rest of the Moon, and to the planets and the asteroids.

Frank and the aliens go watch a little group of settlers emigrating to the asteroid belt. There's three men and three women, each of a different race for maximum genetic diversity. For the high earnestness of their endeavor, they've chosen to take the same last name—Robinson—and they've adopted short first names like labels in a logic diagram: Ali, Ala, Ben, Bea, Cus, and Cis.

The six Robinsons gather in a field near the top of Mount Hamilton by San Jose. Each of them wears an uvvy and holds an alla. They disrobe and stand there nude, preparing to put on some kind of plastic rocket suits.

For this unusual occasion a few journalists are actually present in the flesh. "Why aren't you taking more supplies?" asks a newslady.

"With our uvvies we can get all the software we'll ever need off the Net," says Ala Robinson, a handsome young black woman.

"And the allas can make us our hardware and raw materials from the material of the asteroid," chimes in Cus.

"And our bodies have all the wetware seed cells we'll need," says voluptuous Bea.

"Let's put on our suits," says Ali.

The six pioneers slip on limp piezoplastic suits that are something like mirror-finished wet suits. They put their hands down their sides, the suits grow rigid, and now the six shiny, statuelike forms jet up into the air, vanishing high overhead.

It's not convenient for Frank's saucer to follow the Robinsons, but the aliens are able to patch him into a pay-per-view news feed from the pioneers.

Skipping forward through time, Frank watches as the Robinsons coast along the long spacetime geodesic to the asteroid belt. He's there when they finally land on the asteroid Xerxes, a potato-shaped rock proportioned like a brick that's eight kilometers long. Frank and the aliens watch from

FIGURE 45: A Piezoplastic Rocket Suit

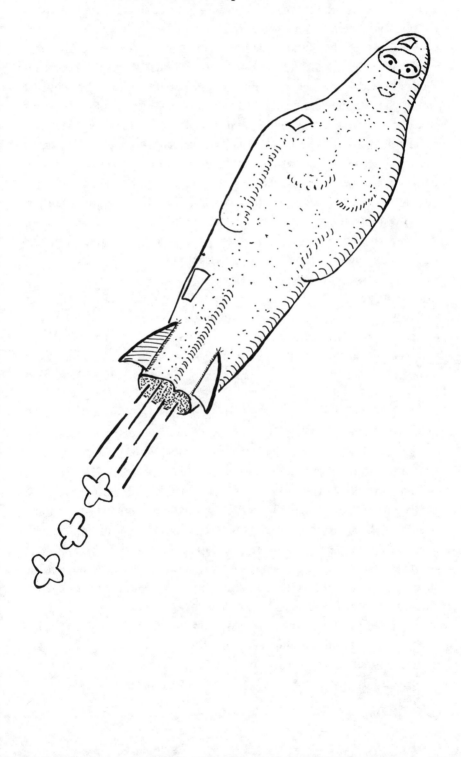

Bea's point of view as the Robinsons use their allas to tunnel into the interior of Xerxes. Rather than using any kind of cameras, the Robinsons simply send direct radiotelepathic feeds of what their own eyes see.

In a few months, Xerxes is hollowed out like an insect's egg case, with a kilometer-thick hull, and an ellipsoidal internal space six kilometers long. A condition of near weightlessness prevails within the immense stone space station that the asteroid has become—although there is a very weak gravitational effect tending to make things drift to the end caps.

Through clever use of the alla, the last half-kilometer of space at each end of the hollow potato has been partitioned off with stone latticework, and beyond these lattices lie two great, wobbling internal seas. And all over the potato's interior, scattered little bulging ponds are fastened to the walls with domes of golden mesh.

The vast space's interior is lit by great polka-dot portholes of ten-meter-thick diamond that the alla has set into holes in the asteroidal rind. In the process of casting off the extra mass from the asteroid's interior, the Robinsons have replaced the asteroid's original random tumbling with a very slow rotation along its long axis—like a chicken on a spit. To give a familiar sense of time, the rotation speed is tuned to one complete cycle per twenty-four hours. As the spotted asteroid turns, beams of brightness play about the interior like shafts of sunlight from a cloud-dappled sky.

Much of the remaining outer surface of Xerxes is covered with photoelectric materials; the Robinsons have done a bootstrapping process of using their allas to make photocells, using the photocells' energy to recharge the allas, making still more photocells, and so on.

The air within Xerxes is sweet and clear, the women are pregnant, yet the environment is blandly sterile, for there are no plants and animals other than the six Robinsons and whatever minute life-forms inhabit their bodies. So now they get to work populating their world.

They use the allas to create some wetware engineering equipment, such as cell-surgery pipettes and amniotic baths. And then they use the allas to instantiate a number of DNA maps that they download via the uvvy—DNA for fruits, vines, insects, lizards, birds, and the like. The alla-created DNA strands are implanted into live cells scraped off the insides of the Robinsons' lips—each of the settlers is eager to have some of his or her cells used, eager to be an Adam or an Eve for this new world.

FIGURE 46: Eden in a Hollow Asteroid

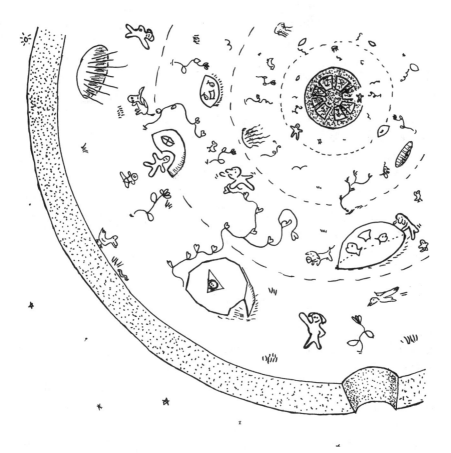

Within two years, the interior of Xerxes is a complete Eden, with great flowering vines, flapping birds, fish in the cap seas, and bioengineered pod homes for the growing families Robinson.

When the Robinsons need something from Earth—such as new model allas, or more uvvy time—they pay off the Earthlings by capturing small asteroids and launching them into orbit near Earth. One natural resource still needed in this new femtotechnological age is raw matter, and Earth's environmentalists don't feel good about alla-converting entire cubic kilometers of Gaia's substance.

Frank muses upon the simple trinity of things the settlers need besides mass and energy: uvvy for software, alla for hardware, and their own bodies for wetware.

"This is perfection," he says to Steffi. "People could go anywhere like this. If they wanted to, they could even use the asteroid as a multigeneration starship. Aim all the jets one way, and power it out towards Alpha Centauri! Their great-grandchildren could do the landing."

"That would be stupid," says Steffi. "First of all, I've been to Alpha Centauri and there's no planet there you'd like. And even if there was, by the time any fat starship got there, people probably will have traveled there the easy way."

"You mean as personality waves?"

"Right," says Steffi. "The uvvy, alla, and biotech are big discoveries, but they're nothing compared to soul transmission. That's the most important breakthrough we of ♎ yet know of. Why don't we look a little look further, and see if Earthlings will find the way."

3OXING

Joe and Harry eating food they can change. Alla on a thin stick. Strange quark golf ball from an asteroid. It tears off a new thumb.

This dweeb gets a 3ox of a china figurine. Agent lives in a shell on a cliff. People 3oxing antiques. The rich man is bug-zapped into two. How would it feel?

Steffi says she'll put me back at Mount Rushmore, hooray!

Lulu figures out matter maps. Her alla's a magic wand now: a baby in the shuffle-board court. I'm getting overloaded.

Strange Matter

The saucer goes into the lab of Femtotechnology Unlimited to watch Harry and Joe. They're eating a meal of alla-made spaghetti, preternaturally smooth tubes of pasta topped by mathematically perfect polyhedra of shimmering tofu. They're drinking out of water glasses that never get empty, a bit like Frank's alien water mug.

Joe sets down his fork with his plate still half-full. He takes out a little alla that's mounted on the tip of a stick like a wand. He waves his alla over his plate, turning the leftover spaghetti into ice cream. And then he uses the alla to turn his fork into a spoon.

"Actually, this food sucks," Joe says to Harry. "It's too smooth. Isn't there some way to put in more texture—like natural things have?"

"Natural things are fractals," says Harry. "They have layer upon layer of detail. The alla can generate *random* fractals, but organic things have more structure than that. It's too hard to write an algorithm to properly represent an organic form. The Holy Grail would be to make a living creature with an alla. But so far we have to fake it." Harry's thick lips spread in a smile, and while he's smiling, he reaches his alla across the table and turns Joe's ice cream into something like spinach. "Level four random fractal there, Joe. Good for you."

Joe takes a drink of the water, tastes the spinach, turns it back into ice cream, fills his mouth with water, and squirts the water out of his mouth in a thin stream, letting the alla change the water into air. "Jerk. What we need is a way to directly copy a physical object. Instead of coding it up as a blueprint, you let the object be its own description. Like making a rubbing of a gravestone. Or putting ink on a fish and using it to print a picture of its scales. Would there be any way to do that in three dimensions?"

"If we could fold a piece of space over," muses Harry. "And then make the matter pattern soak through. Matter is really just bumps in space, you know. The old *Flatland* thing." Harry draws a picture that Frank copies down.

Frank and the aliens jump ahead a few months to find Joe and Harry in the cafeteria again, still talking.

"So it'll be like three-D Xeroxing," Joe is saying. "We'll call it *three-oxing*. Spelled numeral-3, letter-O, letter-x, *3ox*. Our proprietary trademark."

"I like it," says Harry. "3ox. Fine. Now what's still missing, Joe?"

"I know, I know," says Joe, holding up his alla and turning a spot of air into a tiny little lightbulb that clatters down to the tabletop. "What's missing is how it's gonna work! So how do you fold space over to make a 'rubbing' of an object's bumps?"

"Strange quark matter," says Harry, using his alla to turn the lightbulb into a malted-milk ball, which he eats. "I got a rather large amount of it yesterday."

"Lecture?" says Joe, using his alla to stir carbonation into his water.

Harry grins and begins to talk. "As I've mentioned to you before,

FIGURE 47: Copying a Space Bump by Folding

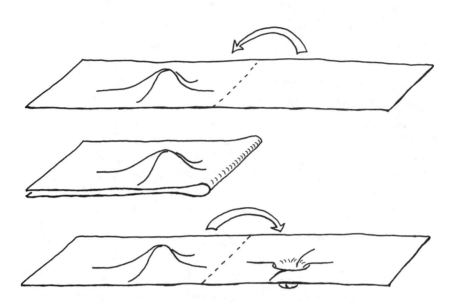

ordinary matter is made of *up* and *down* quarks. There are four other, less common, kinds of quark called the *strange*, the *charm*, the *top*, and the *bottom* quarks.'' With each quark name he flicks his alla and sends a bouncing little polyhedron tumbling across the tabletop. ''Usually you only have three quarks in a quark-bag—remember, a proton is two ups and a down, while a neutron is two downs and an up. But if you throw in other kinds of quarks, you can have as many quarks in the bag as you like. I'm using a mixture of all six kinds. This makes something that's called *strange matter*. A small ball of strange matter is a strangelet, and a star-sized chunk is a strange star. Each of them is just the one quark-bag, you understand.''

''Like a turkey-basting bag,'' says Joe, gingerly cleaning between his teeth with his alla. ''And you're saying you've got the strange matter now?''

''Follow me outside,'' says Harry.

Outside on the grass there is an incredibly dark-looking black disk. The whole lawn seems to sag beneath its weight. The disk is the size of a hockey puck. It has an intense, plutonic aura.

"It's got 10^{39} quarks and a mass of a billion tons," says Harry.
"Where'd you get all that mass?"

"The Robinsons. This used to be an asteroid about one kilometer across. I sent them the blueprints for a quark converter and had them turn one of their neighboring asteroids into strange matter. It's so heavy and dense that it's pushing down with a force of—oh—ten million tons per square inch."

"Why doesn't it sink into the Earth?" asks Joe. "Like the tip of a high heel."

"Ah," says Harry happily. "Good question. The thing is, I stiffened the ground."

"Huh?"

"I injected some fast-growing bucky-fungus into the soil. We're standing on about an acre of the stuff by now. It fills the soil with a mycelium of super-strong bucky tubes. You know what mycelium is, Joe, it's like the white, webby stuff in the ground under a mushroom."

"Gnarly!" Joe steps forward and stretches out his arm, as if to touch the intensely-black disk. "I think I can even feel a little bit of gravitational attraction. Or—ow!" There's a sizzling sound. Joe snatches back his hand, clutching it to his chest. "It zapped me!"

"It's got a very big cloud of electrons," says Harry. "You could think of it as a single, giant atom. And the space distortion around it is—bizarre. I knew you'd try and touch it, Joe. Show me what it did to your hand."

Joe gingerly uncovers his hand, still held cradled against his chest. The hand is red and sunburnt-looking, but no permanent damage seems to be done—except—

"Whoah!" cries Joe. "I've got three thumbs!" He jerks in surprise and the extra thumb drops to the ground.

"So it even works on human flesh," says Harry. "Thanks for testing it, Joe."

"What happened?" Joe is anxiously rubbing his hands and staring down at the thumb on the ground. It's twitching. Joe picks it back up.

"Watch," says Harry. He plucks a yellow dandelion and throws it at the slug of strange matter. There's a high, singing buzz, and two dandelions bounce off of the disk. Joe picks up the flowers and examines them, with Frank and the saucer watching over his shoulder. The blossoms are identical in every respect.

FIGURE 48: Joe's Third Thumb

"The strange matter yanks space over itself like a chilly sleeper tugging a blanket," says Harry. "It makes space take a rubbing of itself, like you said. Put differently, it acts sort of like a beam splitter; like a partially silvered mirror that reflects an image, and also passes an image through. Ordinary matter is a pattern of ripples in space, and the distortion from the strange matter is splitting these ripples in two."

A buzzing honeybee flies at the dark disk, there's a small pop, and two honeybees fly away.

"This isn't exactly a consumer product like the alla," says Joe, chewing the nail of his extra thumb. "Does it have to weigh a billion tons?"

"Yes," says Harry. "In fact, it should weigh a lot more if I'm ever going to 3ox something big. Smaller masses don't affect a big enough region of space. And, you know, even these billion tons aren't going to last forever. Every time it copies something it loses an equivalent amount of mass. And whenever there's a breeze, it's copying air." He looks around. "I'm going to grow a dome over this piece of lawn today and start charging people to use this thing before it melts."

Cloning Collectibles

Frank and the aliens jump a few months further into the fourth millennium, scanning the UV signals for someone talking about using 3ox technology. They hit upon a solitary, yellow-skinned dweeb.

His name is Pol. He has pursed self-indulgent lips like a fat lady, a raggedy beard, long greasy hair, and a white shirt. He is sitting in his home, which is seemingly grown from a fungal puffball. The walls are lined with mushroom-gill shelves, and the shelves are laden with hundreds of little porcelain figurines—Disney characters, Hummel figurines, Kincaid houses—antiques with their colors faded to shades of beige and gray. Pol is examining an ancient paper book on figurines. The illustrations in the book are small and musty, page after page of them. Pol keeps turning back to a picture of a statuette of a bearded little man called Hermit Huck. A Gerard Walloon original. For whatever reason, Pol has got to have this statuette.

He uses his uvvy to call the dealer, a slippery woman named Lona. "Hi, Lona," drones Pol. The simultaneous translation gives him a reedy, monotone voice. "I really must acquire a Hermit Huck. I have three other pieces by Walloon, you know. The Cork-Soaker, the Sock-Tucker, and the Mortar-Forker. Hermit Huck will add a new tone."

"There's only one original Hermit Huck in existence," says Lona. She has hollow cheeks, black lips, and her ears are coated with glued-on diamonds. She lives in a big gold-filigreed barnacle shell on a cliff overlooking the Golden Gate. "Since you're such a good customer, I can get you a deal. With my brokering fee it would come to just under twelve thousand glorks."

"That's—that's impossible," protests Pol. "That's more money than I paid for my whole collection."

"Hermit Huck is one of a kind," repeats Lona. "Would you settle for a reproduction?"

"What do you think I am," splutters Pol. "An amateur?"

"I'm talking about a femtotechnological quark-for-quark 3ox copy," says Lona. "All the weathering, the patina, the glaze chemistry—the whole thing will be there. 3ox is the latest technique. There is absolutely no difference between the copy and the original—in fact, there's no way to tell which is which. In fact, it's debatable if there even *is* a difference. We dealers have agreed, however, to always pick one of the pair and fungus-tag it as being a copy, so that we don't get an explosion of originals. You'll get the fungus-tagged one, of course. But none of your friends needs to know. And, though I wouldn't tell this to anyone but you, Pol, you can remove a fungus-tag with the right kind of retrovirus."

"And the price?"

"I'll have to talk it over with the present Hermit Huck owner. But I think we can agree on something reasonable. Perhaps eighty glorks. And, Pol, you'll be able to recover some of that money by 3oxing off second-generation copies of your Hermit Huck. They'll still be perfect."

"This is going to destroy collecting," says Pol.

"It's happening, baby," says Lona. "So why not get on it early. In fact, I'd like to 3ox your Three Caballeros salt-and-pepper shakers for a client in the East Bay."

"So be it," says Pol. "Where should I meet you?"

"At Femtotechnology Unlimited," says Lona. "They've got the 3ox disks in a bunch of domes there."

Matter Maps

So Frank and the alien saucer follow Lona and Pol back to Femtotechnology Unlimited. There are six domes beside the main building now, each dome bigger than the one before. Smart dog-sized sluggies patrol the grounds, encouraging people to stay in orderly lines. Each person carries some treasured item, and the sluggies direct people to the dome with the proper-sized 3ox disk.

It seems odd to Frank, this far into the future, to see people physically carrying things around. Objects are so atavistic. Some customers have antiques and collectibles that they want 3oxed, others have brought beloved pets, and still others are there to get their children or even themselves copied.

The astounded Frank watches as a wealthy old man leaps at a trillion-ton manhole cover of strange matter. He holds his nose as if he's jumping into a well. The disk makes a sharp crackle like a bug zapper, and two half-as-wealthy old men come bouncing off. Just like Lona said, the transformation happens so smoothly that it's impossible to say which is the original.

"I wonder what that's like for him," says Frank.

"It's very consciousness-expanding," says Steffi. "To have multiple bodies—it makes you realize very clearly that the cosmos is filled with one great Mind, which is God. As it happens, two of my shipmates are identical to me. Lalla? Henny?" Two pieces of rope rear up next to Steffi, and the three of them sway back and forth as one.

"Can we follow that old man home?" asks Frank. The idea of following someone home makes him suddenly remember that he still hasn't asked the aliens to help him escape from the cops. "Oh, but there's something else I need to ask you about too."

"What is that?" says Steffi.

"When we're done finding out about femtotechnology, could you set me back at a different place from where you picked me up?"

"Why? It would be confusing for your fellow Earthlings."

"The cops are after me." He gets out the *Saucer Wisdom* manuscript again. "I stole this from Rudy, you see. I'm afraid that if I go back to the Santa Cruz mountains I'll get arrested before I can patch things up with him."

"You want to return to a different location but on the same planet?"

"Yes. I thought—why not Mount Rushmore? It's in South Dakota. I've always wanted visit there."

"It would cost us quite a bit of energy," says Steffi, after peering into Frank's brain for details. "But I suppose we could do it. Be warned that if we leave you at a different space location, there will be a time

FIGURE 49: Man Getting Twinned

difference as well. It won't be like your other saucer excursions. You won't come back at the same instant where you left.''

"Nobody's going to notice anyway," says Frank, cavalierly assuming that Steffi is talking about an interval equivalent to the time it might take to fly from California to South Dakota. "So that'll be fine. Thank you. Can we follow that old man now?"

"I think I'd rather finish our initial line of investigation," says Steffi primly. "We were trying to find out if humans will fully master femtotechnology."

"This is getting too confused for me," protests Frank. "First there was the alla, and that can make pretty much any kind of simple stuff you want. And then the Robinsons were using software descriptions to create DNA that they put into cells from their mouths that grew into plants and animals. And now there's the 3ox disks and they're like physically cloning people. It's too many different things. And you want to show me more? Maybe you should just take me to Mount Rushmore right now."

"No," says Steffi. "We want to see how this comes out. And you have to help. We're not good at finding the interesting moments without you."

So they skip through the next year or so at Femtotechnology Unlimited, and finally they focus upon a Burmese woman named Lulu Ma who makes the breakthrough discovery. Frank recognizes her from an earlier saucer trip; Herman once took him to Lulu's home lab and he watched Lulu using the alla to make part of a snake.

"It's in the sizzle," Lulu is telling Harry and Joe. She and the two men are on a kind of wooden shuffleboard court in a filigreed alla-made ivory building behind the 3ox domes of Femtotechnology Unlimited. It's an employee recreation area. Each of them has an alla at the end of a long stick, and they're sliding odd things around on the floor.

"When something bounces off the 3ox disks," continues Lulu, "there's a burst of sound and a chirp of what I initially took to be electromagnetic radiation. But it turns out that the chirp has a very extensive Hilbert space spectrum. Although the chirp has a superficial resemblance to, say, a megahertz uvvy signal, if you run it through a Twistermann-Levy dimensional analyzer, you discover that its higher-dimensional components are carrying information in the eight yottahertz range. A yottabyte of information per second, in other words."

"Yottabyte," murmurs Joe. "That's the information capacity of a human body. Septillions."

Lulu grins, waves both hands upwards dramatically, and forges on. "And what kind of information is the chirp carrying? Exactly what one would hope. A matter map. The entire information of the object being duplicated is in the yottahertz Hilbert chirp. It's a holographic matter map, accurate down to the subquark scale. These are my observations. And to cap the discovery, I've found a way to use compression technology to save the chirp as a tidy little file that I can play

FIGURE 50: Lulu and Her Wand

back through my alla. My magic wand.'' Lulu sweeps her alla-ended
shuffleboard stick and a chain of little animals appears—snakes, mice,
birds, lizards.

"The Holy Grail," says Harry softly.

"Anything's possible now," continues Lulu. She poises her alla-
wand, makes a dramatic gesture—and a plump, white baby boy appears
on the floor, kicking and cooing. "I saved his matter map from when a
woman copied him last week. Her name was Maria."

"Oh, wow," says Joe. "Are we stuck with that baby now? Who's
going to take care of him?"

"Don't worry," says Lulu, scooping up the baby and jouncing it cozily. "I told Maria I was going to try making another copy and she said that if it worked, she wanted him."

"This is going to change everything," says Harry.

Mount Rushmore

"He's right," says Steffi. "Though they don't realize it yet, they're on the way to personality waves. First they'll need to figure out how to get the matter-map without having to 3ox the object. And then they'll need to find the way to code the map into a higher-dimensional electromagnetic wave that can reconstitute itself without an alla. So that they can draw the necessary mass right out of the higher-dimensional ether. That's what we do, anyhow. Shall we push on into the future and find out if the humans will succeed? Probably by the mid-4000s would do. Help us tune in on it, Frank."

Frank balks. The thought of going so very far into the future makes him sick with fear. He makes his mind as blank and unhelpful as he can, meanwhile begging at the top of his lungs. "Please take me back! And remember, I can't go back to California, because the cops are after me. Drop me at Mount Rushmore!"

The saucer bucks and yaws, working its way back to the 1990s. Now Frank sees the good old San Jose airport down below, dry and shimmery beneath the beating sun. An airplane hangs in the air near them, static in frozen time.

"Hang on," says Steffi, and all of a sudden the airplane roars towards them, the pilot's face a mask of terror. The saucer—it's a big, bright dragon again—makes a brisk, whizzing noise and takes off for the east, shooting across the High Sierras, the wastelands of Nevada, the flats of Utah, the rolling ranges of Wyoming, and the wooded Black Hills of South Dakota. The landscape flickers stroboscopically as they move; it's hard for Frank to decide if it's day or night. He can't seem to locate the sun. Now they're coming down on Mount Rushmore, a white granite outcrop amidst a pine forest. The park service has built an elaborate reception center near the great carved cliffs. It's definitely daytime all of a sudden, and people are all around, but the people are like frozen statues—because the saucer's switched back to perpendicular time.

The next thing Frank knows he's been set down inside an empty men's room next to the snack bar. He sags against the sink, then starts washing his face. The door opens and another man comes in.

"Excuse me! You forgot to turn the lock."

"That's okay," says Frank. "I'm done in here." He dries his face and exits to the snack bar. There's a great horde of tourists.

Frank picks up a free pamphlet guide to the Black Hills of South Dakota, and gets himself an enormous vanilla ice cream cone. He goes and sits outside, eating his cone and staring at the distant stone heads of the presidents. Washington, Jefferson, Teddy Roosevelt, and Lincoln. The heads aren't as big as he'd expected. What's Teddy Roosevelt doing up there, anyway? Frank formed a lasting dislike for the guy in his tedious grade school history classes.

It's incredible to actually be here at Mount Rushmore. So nice to be out of the saucer. It's a fine, sunny day. Frank figures he'll make his way to the nearest town and get some kind of job. Give Mary a call and see how things are going. Everything will be—and then Frank notices the date on the pamphlet he picked up in the snack bar. June–July, 1996. Two years are gone.

ELEVEN

Notes on Transhumanity

On the morning of Sunday, June 15, 1997, we rose early and headed straight up through the pine woods towards the Devil's Tower. We didn't bother taking a trail, just headed though the woods cross-country. I thought again of the conclusion to *Close Encounters of the Third Kind*— of how the hero makes a beeline for the tower, sure that he'll find a great alien revelation.

By ten in the morning we'd come to the Tower trail that encircles the base of the tower. We happily walked it, marveling at the great, bundled columns of stone. We circled the tower once, twice, three times, talking all the while. By then too many tourists had started to arrive, so Frank and I left the trail to spend the rest of the afternoon in a little clearing near the base of the tower's north side, sitting there picnicking and talking.

One aspect of the transhumanity notes that most readers will find completely unbelievable is Frank's insistence that people will still be reading our *Saucer Wisdom* in the year 4004. It would be nice! In reality, Frank had a fairly transparent motivation for adding this embellishment to his tale. He wanted me to believe that our book was going to make a lot of money, so that I'd be willing to give him a few thousand dollars as an advance on his 20 percent of the book's supposed future earnings. I declined to do so, pointing out that I'd just paid for the plane trip to South Dakota, and that I had a lot of writing and marketing work still ahead of me, with no guaranteed payoff yet in sight.

As well as going over the transhumanity notes and asking me for money, Frank discussed various little corrections to my existing manuscript of *Saucer Wisdom*: The notes on future communication, on aliens, and on future biotechnology. To my relief, he didn't get into arguing with anything I'd written about my encounters with him.

EDITING THE BODY

Helene finally gets sick for a day. I'm alone in the Black Hills High IRC. I fumble the attractor together, the saucer comes. It's three humans from the future! Goola, Perl and Balaam.

I want to go back to 1994 to prevent the break-in, I want to keep Mary. They won't help me. "You have to finish Saucer Wisdom."

They want to tell me about transhumanity. Editing, copying, and transcending the body.

Aug dogs at the juice bar in Rapid. Fingers around a girl's neck. Tails, a shoulder rose that looks like a nudibranch egg ribbon. A Russian sleaze-ball puts a deer-antler bud in a girl's head, she speeds up.

A black farmer with two extra hands, they run along on their fingers next to him. His wife watches him with a Von Dutch eyeball. Back in the house, she's in a tub of amniotic fluid getting rejuvenated.

Their son Rize has turned himself into a merman. Phone up and meet his fiancée. It's lonely to grow old.

The Human Saucerians

After Steffi and the rope aliens drop Frank at Mount Rushmore—the exact date turns out to be June 27, 1996—he makes several efforts to phone up Mary. It takes almost a week to track her down, and meanwhile he's living in a cheap motel near Rapid City, paying his bills with a shit job as the night janitor at a McDonald's.

When Frank finally talks to Mary, he finds out that she's remarried. She doesn't want to get back together with Frank at all. The only good thing is that Mary set aside Frank's share of the money from their joint savings and from the yard sale of their things. She mails Frank a check and he lives a little more comfortably for a few months. He's scared the police are still looking for him in California. He can't think of any particular reason to go anyplace just now, so he stays in Rapid. Autumn is beautiful in South Dakota.

When the winter comes on, Frank gets a job at the instructional resources center at Black Hills High School, just north of Rapid. His boss at the IRC is a skinny woman named Helene Lundy. She has lank, colorless hair, and her mouth becomes a disapproving O whenever sex is mentioned in her presence, which is not very often. Her pale skin is pink around the nostrils from a perpetual case of the sniffles.

One of Frank's reasons for taking the IRC job is that he hopes to use their equipment to contact the aliens. He's in a state of denial over losing Mary; he feels that if he can just get back in touch with the aliens, his familiar old life will somehow start up again.

But Helene Lundy is always there in the equipment room, day in and day out, keeping a strict eye on everything, softly blowing her nose into wadded little tissues. Finally on Wednesday, April 9, 1997, Helene fails to report for work. This is Frank's big chance.

He locks up the IRC for lunch hour and stays inside. The voices and footsteps of the students echo through the long, waxy high school halls, a tender, nostalgic sound that makes Frank ache for his past. Frank makes sure the blinds on all the windows are pulled down, and then he arranges three cameras and TV monitors. The equipment is unfamiliar and Frank's hands are trembling. It takes him much longer than usual to tune in the right kind of pattern, but finally he gets the proper tangle of bright, smeared blobs and lines.

A saucer moves out of the TV screens and envelops Frank. He groans with joy. Unlike the other times, the masters of this saucer rotate themselves into Frank's field of view without even being asked. To Frank's great surprise, they're humans, or rather transhumans: Our descendants from the fifth millennium.

There are three of them, a man, a woman, and an androgyne. In physical form they're very handsome, almost godlike. The woman, named Goola, is dark-skinned and has flowing blond hair. The man is Perl; he is an Apollo with strong, green teeth. The third is Balaam, a winsome, plump figure with a sweet round face and an acid tongue.

"How now, Frank Shook," says plump, fey, butch Balaam, seamlessly translating her/his words into an odd, campy dialect. "You look to be even more of a schnauzer than I expected. We're here because of your little book. 'Tis our fate to make the prophecies come true. *Mektoub*, as the Arabs say. *It is written.* A cliché, but so is life. I just flew in from California, and boy are my wings tired.''

Rather than responding to any of this, Frank immediately hollers out his big request. He's been rehearsing it to himself for the past eight months. ''Take me back to eleven P.M., June thirtieth, 1994! Rudy Rucker's house in Los Perros! If I get there in time, I can stop myself from breaking in! And then I won't have to run off to South Dakota and lose two years. I'll start up my old life with Mary.''

''Assume we were willing to help you,'' says the inhumanly handsome Perl, who speaks like a British actor impersonating a scientist. ''If we helped you change your past, then you *wouldn't* be here now. And if you weren't here now, then we certainly wouldn't be helping you. And in that case, your past wouldn't change and you *would* be here now. So if we're willing to help you, then you're here if and only if you're not here. A contradiction. The universe doesn't allow contradictions. Therefore, the initial assumption is false. Meaning that we're not going to help you. Q.E.D. and a priori. Logic is very powerful.'' Perl's green teeth flash in a satisfied smile.

''How's the universe going to know if there's a contradiction?'' protests Frank. ''Maybe it's not paying attention.''

''The cosmos is God's bod,'' says dark-skinned Goola, tossing her long, blond hair. Her mode of discourse is hipster-ecstatic. ''All space and all time. Each moment is tweaked to make the greatest harmony across the spacetime dimensions. Reality is like the smooth plastic wrapper around a basket filled with gnarly fruits. Like a silk sheet over a bed filled with phantom lovers. Like the final telling of a perfect tale.''

''But—what happens if you try and *force* a contradiction?'' insists Frank.

''Nothing,'' says Balaam a little impatiently, ''Because, my dear, *no can do*. What's written is the holy writ. We're going to fill your little brain up with transhumanity and then you're going to finish the notes for *Saucer Wisdom*.''

''You've read it?'' exclaims Frank.

''Oh, yes,'' says Perl, warming up. ''It's quite the classic, on a par with Plato's *Dialogues*. I'm dead chuffed to meet you. It's like meeting Socrates.''

"Yeah?" says Frank. "Yeah? Well, what if I decide to cause a contradiction by changing my book's name, huh? Teach you a lesson. I'll call it *Frank Shook's Alien Visions of the Coming Millennia.*"

"What a tacky, tacky name," titters Balaam, leaning his/her head on Perl's shoulder. "Rudy Rucker has far better taste than that." S/he smirks and shrugs. "You can't change the future any more than you can change the past, Frank Shook. People don't do the opposite of what they know they're going to do."

"What about this," presses Frank, still trying to find a way back to his past. "What if you take me back to June thirtieth, 1994—and then I branch into a different universe! That way there wouldn't have to be a contradiction. Oh, please take me back. I can be with Mary in an alternate world."

"There's only one universe," says Perl. "If time branched, then nothing would matter. Everything would always happen no matter what anyone did. Instead of a single, wonderfully thought-out line, time would be a stupid pudding of anyhow outcomes."

"But I saw alternate futures when I was previewing my meeting with Rudy at the Jahva House," protests Frank. "Weren't those real?"

"Those were virtual futures," says Perl, idly fondling Balaam. "Mental constructs, like dreams. But only *one* of those futures actually happened. If *everything* happened, no work would be accomplished by our living out this particular cosmos. There wouldn't be any reason for us to exist."

"I don't understand," says Frank.

"The world is a cosmic work of art," says Goola, running her hands over her body. "Like a sculpture from a block of marble. If all the worlds existed, you'd just have the uncarved block—the same as if no worlds existed. Everything and Nothing are easy; it's what's in-between that's hard. There's only one big *aha*, and we're it."

Frank still doesn't understand, but he knows that he's smitten by Goola. He reaches out towards her and—

"Let's be on our merry, fairy way," interrupts Balaam. "We'll give Frank an inspirational looky-look at the highlights of transhumanity, drop him back to gross out Helene Lundy, and then we're for the galactic beyond!"

The saucer rises up out of Black Hills High School, up into the gusty, wet April sky. A thaw is on; melting snow is everywhere. To the south lies Rapid City and the Black Hills, their dark pines emerging from the winter's frosting. In the other directions the soggy prairie rolls as far as the eye can see, patches of Easter-green grass peeking through the grayish old snow. The saucer darts towards Rapid's meager sprawl.

"What's transhumanity?" asks Frank.

"Editing the body, copying the body, transcending the body," says Perl, extending a perfect finger for each one. "We're going to show you all about it so that you can write it up. And then people will know the way."

"I'll be your guide to body editing," says Balaam. "It's done *me* a world of good."

"And I, as the scion of a long line of clones, will show you how body copying is done," says Perl.

"Transcendence is *my* thing," says Goola promisingly. "So I'll be dessert."

"He looks confused," says Balaam. "Let me make it country simple. Transhumanity means being like us! We never get sick, for instance. What's that like, anyway, Frank? It must be so pukeful to sneeze uncontrollably. A whole chain of them? Does that ever happen to you?"

"Sure," says Frank. "There's a tingling in your nose and then kind of an explosion. It's really bad for you to try and hold in a sneeze. I did that once when I was bending over, and it threw a crick into my back that lasted six weeks."

"You poor man," says Goola. "I think you're so brave." She keeps smiling at Frank, and he's trying not to stare at her too hard.

"Find our first stop, Balaam," says Perl.

"I'm scanning now," says Balaam, her/his playful eyes glazing over. "I'm looking for the local aug dogs. Yes, by the late twenty-third century they'll be here in Rapid too. We'll just scan ahead and—ooh, this scanning through time is a gas, isn't it, Goola?"

Goola bends over as if too see better, which places her inviting ass right in Frank's face. Frank can feel the warmth of Goola's body, can smell the scent of her skin.

Beyond Goola, the view from the saucer is gray and indistinct, but now Balaam makes an abrupt turn and things snap into focus. Outside it's a Rapid City summer's eve, 2297.

"Now we channel the scene at the High Plains Smoothie stand," says Balaam. "I'll interpret the people's speech into Frank's kind of English. What with me being such the thespian."

The saucer shrinks, grows transparent, draws closer to the juice stand.

Aug Dogs

As a building, High Plains Smoothie is a biotechnological marvel. Its exterior is a twenty-foot-high transparent dome, organically grown. Frank and the saucerians fly in through one of the dome's high, arched doorways. Within the dome, seven tough-looking tree trunks rise up to form an umbrella of linked leafy branches.

The room is filled with teenagers, in some ways not so different from the kids Frank's used to seeing at Black Hills High. They're clothed in tank-grown leathers and intricately patterned loom-plant cloths, with belts and vests of flickering piezoplastic. What really sets these twenty-third-century youths apart is their extensive body augmentations.

"They're 'aug dogs,' " explains Goola, her breath hot against Frank's cheek. "A slang expression, you wave. This is the Rapid aug dog hangout."

Fruits of all kinds hang up in the branches: strawberries, guavas, peaches, bananas, pineapples, oranges, raspberries, kiwis, and more. Each fruit is about the size of an apple; that is, the raspberries are relatively large, while the pineapples are accordingly small.

One girl is evidently the stand's manager. Though she mingles with the customers, she stands apart because she wears a starched-white jacket, left open so as to show her breasts. The others call her Durleen. She wears a necklace of—fingers.

Upon closer inspection, Frank sees that it's not actually a necklace, it's a permanent body modification. There are thirteen fingers growing out of the base of Durleen's neck, arranged like spikes in a dog collar, all pointing out, each finger with its own meticulously painted fingernail.

The fingers curl and gesture as Durleen animatedly talks with a boy named Stig, a pale blond fellow who looks quite normal until Frank notices that he's leaning back on what would seem to be thin air. When the saucer circles around behind the boy, Frank sees that Stig has a great,

FIGURE 51: Durleen's Neck Fingers

thick lizard tail growing out through a hole in the back of his scaly leather jeans.

"My Jena's gonna plant an antler bud by one of her ears tonight!" Durleen is telling Stig.

"For true?" says Stig. "Me, I'd roar some pig tusks." He touches his jaw. "Snarfy!"

"Wouldn't match your tail, Stig," says a boy named Junit. Junit's skin has a shiny, liverish sheen and is dappled with changeable spots. "I say always aug to one plan." There are suckers on the backs of Junit's hands, and his silky, sea-green shirt's openings show scattered little tentacles on his hairless chest.

"Oh, don't be so desult," says the girl named Jena. She's small and blond, with round, gray eyes. "Random has rhythm." She has a flesh-colored rose attached to one shoulder, an extra tongue in the hollow of her neck, a pair of long, lobsterlike feelers growing out of the left side of her head, a stout penis dangling from one side of her bare midriff, and what look like insect mandibles on the backs of some her knuckles. She laughs knowingly and gives one of Durleen's neck-fingers a butterfly kiss.

An older man named Volga asks Durleen for a kiwi juice. He has a heavy Russian accent and seems a bit stoned. Durleen pulls something that looks like a condom out of her coat pocket, then reaches up to stretch the piezoplastic sac over one of the dangling kiwi fruits. The piezoplastic makes quick, savage kneading motions, then drops down from the tree into Durleen's hands. She gives it to Volga, who wetly sucks the pulped juice from the sac. Overhead, a small kiwi has already formed at the old one's spot, and is visibly growing to the standard size.

"How do the fruits come back so fast?" wonders Jena.

"*Gibberlin*," says Volga. "Is an all-purpose growth hormone derived from animal steroids and plant gibberellins. Is speeding up the biological clock. That's vhat I use for my augs as vell. Are you ready for zat antler now, Jena?"

"Sure," says brave little Jena.

Volga finishes his juice and tosses the empty husk to Durleen. And then he produces a bloody little gobbet of something—the antler bud.

"From a deer I vas killing last night. Lie yourself down, Jena."

Jena lays her sweet young head down on a couch. Volga produces a protean piezoplastic tool that seems to convert itself into whatever he needs. It starts as a power nozzle that sprays an anesthetic mist above Jena's ear, becomes a scalpel that snips away a disk of her skin, and then mutates into an abrasive wheel that grinds a little depression into Jena's bared skull.

Volga sets the antler bud into place, patches the hole with the disk of Jena's skin, and smoothes on a healing ointment. His hands seem clumsy and shaky; it's painful to watch him work.

"How fast will it grow?" murmurs Jena.

"This is depending on how much gibberlin I use," says Volga, converting his little tool back into a spray nozzle.

"Use a lot," says Jena. "I can't wait forever."

As Volga is misting in the gibberlin, his fingers slip and he loses control of the nozzle. A tight plume of gibberlin shoots directly into Jena's nose.

"Hoo boy," says Volga, turning the spray off. But it's too late. The gibberlin is in Jena's bloodstream and she's living in fast-forward. The antler shoots out of her head like a tulip in a stop-action nature movie.

Jena dashes around the High Plains Smoothie so fast that the air grows warm. One by one she tears down every single fruit and gobbles them, she's living out weeks and months in just a few minutes. Finally the stuff wears off and she collapses, weak with hunger, her hair and fingernails preternaturally long. The antler curls up over her head like a bonnet. Durleen falls to her knees and cradles Jena in her arms.

"Whew," says Jena, catching her breath. "Chill out, Durleen, I'm perfectly okay. I'm just glad I was wearing my uvvy for all that time. It felt like six months. I did all my homework for the rest of the year. But that could have killed me if I hadn't been in here with all this food. Stupid Volga has to pay for it, not me."

As Stig raises his great lizard tail and moves menacingly towards Volga, the little saucer darts back up into the sky.

Archipelago People

"We'll be able to pick up on not one, not two, but *three* different transhumanity trips with the Embrey family," says Balaam, his/her eyes rolled partway back into her/his head as s/he studies the incoming mega-

channel UV scan. These saucerians don't need Frank's help in win-
nowing through human data. "Archipelago people, rejuvenation, and
mermen."

"The highly optimal Balaam," says Perl, admiringly.

"It helps to have read *Saucer Wisdom*," puts in Goola, just a bit
snippily. There's some undercurrents of jealousy in this ménage à trois.

The surroundings go briefly gray, and then the saucer is angling
down to a narrow gravel road. It's just after dawn on winter's day. They
follow the icy road around a corner, take a driveway into a barnyard of
rutted drifts, and catch up with a black farmer named Otis Embrey. The
rising sun gilds the snow with orange and turns the shadows purple. The
sky is a crystalline blue. It's a beautiful morning.

Otis has an expressive, thoughtful face—big eyes and a wide mouth
with calm, thick lips. Vertical-smile wrinkles and patchy black stubble
on his chin. He's wearing fluorescent-orange quilted overalls. Two small,
dark animals follow him across the barnyard, scampering along in the
fresh footsteps he leaves in the crunchy snow. A third little animal flies
through the air, circling Otis's head.

The saucer zooms closer, and now Frank can see that the "animals"
on the ground are two, strong black hands, running along on their fingertips. Each hand ends with a rounded knot of muscle at its wrist. Set
into the knots of muscle are small, shiny eyes. As for the thing that's
flapping in the air: It's an eyeball with a pair of bat wings. The three
curious little creatures wear uvvy patches of piezoplastic, as does Otis.

"Cow, cow, cow!" calls out Otis, stepping into the barn. "Good
mur-ning to you!"

Recognizing Otis's rich, cheerful voice, the cattle in the barn raise
their heads and moo. There's six cows and a bull. Otis sets to work
filling their mangers. Meanwhile, his two extra hands milk the cows.
Frank notices that the cows' milk is stored in thermos bottles, each cow's
bottle with a different kind of label.

"They're pharmaceutical cows," explains Perl, sensing Frank's un-
spoken question.

"And what about the flying eyeball?" asks Frank.

"That's from his wife, Camilla," says Goola.

"We'll do a peekaboo on her," says Balaam.

FIGURE 52: Otis's Extra Hands

The NuYu Do-It-Yourself Fourteen-Day Rejuvenation Bath
The saucer skips a half hour further in time. Otis is in the farmhouse kitchen making himself some toast and scrambled eggs. His extra hands are resting by the sink, rubbing Bag Balm Ointment into themselves. The eyeball perches on the back of a chair, watching Otis.

"Let's listen," says Perl, and they tune in on the uvvy conversation between Otis and the eyeball.

"Only three more days," the eyeball is saying. It has a woman's gentle voice.

"I'll be glad to have you back," Otis answers through his uvvy. "Mighty cold in bed by myself these long winter nights." He sets his plate down on the table and begins eating. "You feelin' okay today, Camilla?"

"I—I guess so. I looked at myself this morning and it scared me. I look like a skinned rabbit. At least my joints and organs are done, and my muscles are back in place. Thank goodness I can't feel anything."

"Me, I sure wouldn't go through all this mess," says Otis.

"But I'll be so healthy and so—beautiful."

"Hope you don't end up too good-lookin' for me," says Otis. "Hope you don't go out and find you a younger man."

"You still look good to me, darling."

"And you looked good to me just the way you was befo'. It's that Rize put this crazy idea in your head. Our son the fish." Otis finishes his eggs and pours himself a fresh cup of coffee.

"I thought maybe we could try and call him today," says the woman's voice. "After you tend to me."

"I'd like that," says Otis. "I miss the boy." He gets to his feet and puts his dishes in the sink. His extra hands set to work washing them.

"Hang on, Mama, 'cause here I come," calls Otis as he starts up the staircase to the second floor. The farmhouse is an old-fashioned human-built structure with creaky wooden steps. The eyeball flutters along in Otis's wake as he heads down the hall and into the bathroom.

The tub is filled with a straw-colored liquid that reminds Frank of automotive transmission fluid. Pale winter light slants in through the window, lighting up the depths of the tub. Submerged there lies the form of a woman, a woman with no skin, a flayed figure of sinew and muscle. The eyeball alights on the shower rod and sits over the tub.

"Whew," says Otis, averting his gaze. "Sho' hope this works, Camilla. Guess I'll fire up today's instructions."

He accesses a program stored in his uvvy, and a to-do list seems to appear in the air before him. Hooked in as he is, Frank can see it too.

"Day eleven," reads Otis. "Step one. Replenish amniotic fluid with six liters of lukewarm water and one envelope of NuYu SkinGro."

There's a large cardboard box of NuYu supplies sitting on the floor by the tub. The box has a NuYu logo on it, which is a picture of a smiling sun.

FIGURE 53: The Tub and the Eyeball

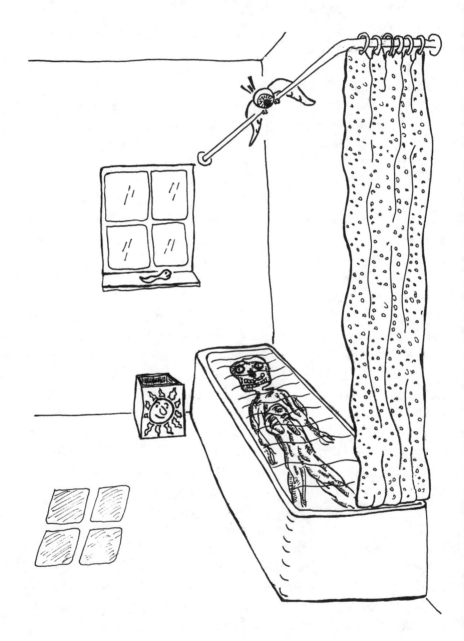

"Only white people could of made up something this dumb," mutters Otis as he measures out the water and tears open a wax-paper envelope of powder. The envelope has a NuYu logo on it too. "I hope you still a black woman when we done with this, Camilla." He sprinkles the powder into the tub.

"Of course I will be," says Camilla. "I specifically ordered light-bronze skin. You helped me pick it out. Doesn't it say so on the envelope? Hold it up." The eyeball leans forward to peer. "You see? Says *light bronze* right there."

"Well, it might turn out like when you pick out a can of paint and it never look the same as the sample," grumbles Otis, discarding the empty envelope. "Be just like the Man to try and turn everybody white."

"You're feeling ornery today, aren't you?" says Camilla.

"Yeah, I guess I am," says Otis, sighing. "Gettin' lonely." He reads again from the floating list. "Step two. Place the NuYu Aesthetician Sluggie into the amniotic bath." He walks over to the window and picks up a sluggie that's sunning itself on the sill. "You got work today," he tells the sluggie, and drops it into the tub. It writhes and stretches itself before settling onto the bare muscles of Camilla's belly.

"Step three," reads Otis. "Wait eight hours, remove the NuYu Aesthetician Sluggie from the tub, rinse it off, and place it in a warm, well-lit place." He ventures one more glance into the tub, where the NuYu Aesthetician Sluggie is very busy. Otis shakes his head and turns off the floating list. "We done for now, Camilla. Let's go back down in the kitchen and give that boy Rize a call."

Mermen

Frank and the saucerians follow Otis, and Camilla's remote eye, back downstairs. Otis starts trying to make an uvvy-call to Rize, but he's not getting through. He goes outside to scrape the snow and ice off his special antenna dish. His extra hands help him; for this intensive task they're wearing leather gloves.

The day is spoiling; the wind has picked up and the sun is behind clouds. Otis finishes with the antenna and hurries back into the kitchen, cozy with a wood fire burning in an old metal stove. Now at last Otis and Camilla make contact. Frank and the saucerians listen in.

Rize is a merman; they see through his eyes. He's looking at a mermaid who swims by his side. Her finned legs are separate, not fused into a fishtail like the traditional mermaid image. Her face is green and pretty. There are gill slits in her neck.

Rize looks down at his own green body, at his finned legs and webbed hands. His eyes can see remarkably far down into the deep. There are fish below, and a school of small, silvery squid.

"Hello, Rize," calls Camilla. "How are you?"

"Fine, Mom. How's your rejuvenation going?"

"Tolerable. I'm tired of thinking about it. Tell me what you've been up to."

"Lots of work, lots of play," says Rize. "Look back there at the development I've been working on." Rize turns his head, and Frank can see scores of glowing, round Sea Homes hovering like Christmas ornaments.

"I miss you, boy," says Otis. "Who's your friend?"

"This is Silla," says Rize. "We're going to get married next month!"

"Is she black?" Otis wants to know.

"She's green, Dad," says Rize. "Green like me."

"Hello, Mr. and Mrs. Embrey," says Silla. "It's nice to meet you."

"Are you in construction too?" asks Camilla.

"I talk to the dolphins," says Silla. She makes a sharp chirping noise and two bottlenose dolphins appear at her side. "They help us find good places to anchor the house-kelp holdfasts."

"You can understand them?" marvels Camilla.

"A little bit," says Silla. "I have part of a dolphin's brain grafted into mine. The hearing lobes. Dolphins can see by sound, you know. And since their faces don't show much expression, they also use their sound patterns to show emotion." Silla opens her mouth and lets out a prolonged twitter. The grinning dolphins nod vigorously.

"They're saying hello to you, Mom and Dad," says Rize. "They're going to come to our wedding."

"Well, I 'spect we'll be there too," says Otis.

"It's going to be underwater," says Rize. "But I can get you some really nice piezoplastic dive suits."

FIGURE 54: Silla the Mermaid

"We can hardly wait," says Camilla.

They talk a little more, and then the connection breaks. Otis sits in the winter kitchen talking things over with his wife's remote eyeball. They're happy about the coming wedding and the chance to see their faraway son.

COPYING THE BODY

A sick old man like a medieval painting of "The Death of the Miser." Mr. Sloane. He doesn't want anyone to know he's dying. They use a life-box to put his mind in a clone. And squeeze juice out of his brain. The clone walks!

I think Goola likes me. I wonder if I have a chance at her. It's been such a long time since I had a woman.

Boba Shekk is a guy like a screenwriter. 26th C. He can't keep up. He gets two ohmies, these are better clones than Mr. Sloane had in the 24th C. Mark and John. Boba programs them by sneezing all over them. John takes Boba's place.

The Ang Ous. Funhouse mirror of the thousand minds. The people's will vs. the kamikaze clones. Turn them into space bugs and shoot them into orbit.

"In another couple of centuries after the Embreys, the human race masters a primitive technology for copying one's body and mind," says Perl. "It's something I'm particularly interested in, so now I'm taking over. Good job, by the way, Balaam."

"Thanks," says Balaam. "You're going to show him Shekk's ohmies?"

"Spot on," says Perl. "But first we'll watch an old-fashioned brain-juice extraction."

"Things were so wild and shaggy in ancient times," exclaims Goola. "So *whatever*. Can you imagine driving an automobile? Oh, I bet Frank's done it! Is it mad and wonderful?"

"Well, it takes a steady hand," brags Frank, basking in Goola's warm attention. "I'm considered quite a good driver. I wish you could come for a ride in the van I just bought, Goola. Solid as a rock at eighty-five miles per."

"Oooh," says Goola, her eyes sparkling. "And with the other cars careening by only inches away! Full of drunk drivers? How exhilarating.

And all the stagecoaches? Have you ever been ambushed by Native Americans?''

''Just last month some kids outside an Indian gambling casino asked me for five bucks for watching my van,'' says Wild West Frank. ''But it was no big thing. I was ahead seventy bucks from the craps table.'' Goola is sitting down next to him now, all ears.

''Are you going to be able to find a good brain-juice scene in Rapid City, Perl?'' asks Balaam.

Perl quietly tries for a minute, scanning away, then suddenly slaps his forehead. ''Right-o, there aren't any here! Don't you remember from *Saucer Wisdom*? We take Frank back to California for the rest of his tour. Rapid City doesn't have the things we need to show him.''

''Hold on there!'' cries Frank. ''I don't want to have any more missing time!''

''Not to worry,'' says Perl. ''If we bring you back here to Rapid when we're done, it'll all come out roses.''

''You won't be missing but a nanosecond, Frankie-pie,'' adds Balaam. ''We're doing you a big favor burning up all those glorks on you. *Mektoub*, isn't it?''

Sloane's Clone

The saucer makes a curious roaring sound and tears across the Western plains, the sky flickering from day to night to day many hundreds of times a minute. Though Frank's not sure of this, it looks as if the hurtling Sun is sweeping across the skies from west to east. Which seems good, because this ought to mean an accumulation of extra time to make up for the missing time that the trip back to Rapid will produce. Frank starts to ask about this, but Goola forestalls his question with a reassuring pat on the leg.

Below them is Russian Hill in San Francisco. It's a stormy winter night of gusty wind and pouring rain.

The saucer homes in on a great castlelike house grown from fine, rare woods. In a vaulted bedroom lies a frail old man in a great ebony bed. The room is lit by candles; there are thick red-velvet curtains and mirrors in golden frames.

Sitting by the dying man's bed is a trim little woman with a tight blond bun. She wears a gray flannel skirt and jacket. Resting on a delicately lacquered table beside her is a small, glowing red cube.

"Can you hear me, Mr. Sloane?" asks the woman.

"Yyesss, Carol," wheezes the old man. His fingers are picking distractedly at the sheet. Frank recalls having heard somewhere that dying people do this, and that there's even a special word for it, which he can't remember.

Balaam senses Frank's wondering about the word and murmurs it to him. "It's called *floccilation*, Frankie. Mr. Sloane is *floccilating*."

The saucer moves around to get a better view of the dying miser. His nose and Adam's apple protrude hugely. His toothless mouth gapes, sucking in air. And attached to the back of his neck is a piezoplastic uvvy.

"We've made a very complete lifebox file for you now, Mr. Sloane," says Carol. "It's right here in the S-cube. You can rest easy about passing on."

"I can't see," whispers Mr. Sloane. "The light is too bright."

"I'll put out another candle," says Carol, pinching out a flame that's within reach. "You'll be happy to know that your new body is almost ready."

The old man's eyelids flutter and his mouth twitches. "Not ready yet? Body not ready?" His fingers continue to pluck at the sheets.

"To do it properly we can't grow it any faster," says Carol. "Otherwise there might be anomalies. Even with gibberlin, a top-quality clone still takes a year to grow to adult size. But as soon as it's ready we'll use your lifebox file to program you right in. You'll only be—out of circulation—for a short time."

"Don't tuh-tell," murmurs Mr. Sloane, and dies.

Carol leans over him, holds a mirror to his mouth, pulls the sheet over his face. She stands, picks up the glowing red S-cube, and walks next door to a laboratory room that is as futuristic as Mr. Sloane's bedchamber was medieval.

The most prominent feature of the room is a long, glass tank in which floats the partially formed figure of a young man. Two high-cheekboned, burr-cut Vietnamese men are in the lab, already on their feet as Carol enters.

"It's time?" says one.

"Yes, Dieu," says Carol. "Quick, you and Thieu get the body and cut it up."

Dieu and Thieu hustle into the bedroom and carry Mr. Sloane out on a stretcher. To spare Frank and themselves the sight of the corpse's dissection, the saucerians jump an hour forward in time.

Now Dieu and Thieu are using the corpse's tissues to help seed and nourish the clone in the tank. And Carol has the dead man's brain in a sluggie sack, which is distilling out memory molecules. Three little tumorlike bumps on the sides of the sack are accumulating the refined juices.

As they work, there is a ringing sound and the face of a young man appears on a view screen. "May I speak to my father?" His voice is cold and snobby.

"Of course, Mr. Sloane," says Carol, with a knowing wink at Dieu. She patches the S-cube into the uvvy call, and now Mr. Sloane's life-box begins carrying on a conversation with his son.

"How are you feeling, Dad?" says the son. "I wish you'd let me come by and visit."

"I feel like shit, Brett. As you very well know. But I could still bounce back. You'd hate that, wouldn't you? No inheritance yet."

"You know you're dying, old fool," snaps young Brett Sloane. "Your disease is incurable. Just get this over with so my real life can begin."

"Keep waiting," says the S-cube, and ends the call.

"This is a real soap opera," says Frank.

"It gets better," says Balaam, and jumps them five months into the future.

Dieu, Thieu, and Carol have taken the clone out of its tank and laid it on a table. It's all finished! With some effort, Frank can make out a resemblance between the clone's face and the features of the dying old man. The clone is breathing slowly and deeply, like an unconscious person.

At Carol's touch, the sluggie sack grows a long, sharp, intimidating needle. Sticking out her tongue a little as she focuses on the delicate task, Carol works the needle in through the sutures of the clone's skull, injecting brain-memory juices into three different locations.

Next, Carol attaches the S-cube to a large, hoodlike uvvy, which she slips over the clone's head. The seconds tick by. Carol does something else to the S-cube, and now a twitching starts down in the clone's toes and travels up his body. The clone's legs tremble as if marching in place; his fingers began to flutter. Another couple of seconds elapse, and now the clone abruptly sits up and pulls the uvvy hood off his head.

FIGURE 55: A Sluggie Sack Brain Injection

"I'm back!" exults the clone. The timbre of his voice is just the same as the old man's had been.

"Do you want to tell Brett the bad news?" says Carol.

"I—" The Sloane clone pauses, staring off into the air. "I've just been with God, Carol."

"Sir?"

"God is love. And I'm alive again. Why should I further torture my poor son?"

"Mr. Sloane," puts in Dieu, "if you don't mind my saying so, your son Brett is a jerk. He doesn't deserve anything from you."

"Yes and no, Dieu," says the clone. "Yes, he's a jerk, but he *exactly* deserves everything I have. Maybe it's a reward—or maybe it's a punishment. Just go ahead and tell him old Mr. Sloane is dead. Me—I'm taking off to bum around the world. I don't need any money. I'm young and strong, with a lifetime's wisdom. More than a lifetime."

The Sloane clone pulls on the clothes they've laid out for him and walks across the lab. He stops at the door and glances back. "Carol. Want to come along?"

"Cut!" chuckles Balaam and skips out into paratime.

Shekk's Ohmies

"As you just saw," says Perl, "the twenty-fourth century cloning process had certain imperfections. One, a person had to die in order to get their memory molecules properly extracted, and, two, even then the clone's behavior was not reliably similar to the anticipated outcome. By the twenty-sixth century the situation was somewhat improved. Inevitably the lifebox technology got better. And people found a fairly crude biotech method of synthesizing the memory molecules. Creating programmed clones of oneself became quite the rage. People called the programmed clones *ohmies*, though I'm not sure why."

"It stood for *other me*," says Goola. "Perl's going to show us some of the ohmies right now, Frank. I've always heard they weren't very vibrant. And that the second- and higher-level ohmies were even more zombielike."

"Well, I'm not surprised," says Balaam. "Those prefemtotechnology times were so squalid, so barbaric. Have you found Shekk yet, Perl?"

While Perl scans, Goola is busy leaning over Frank's lap to look at his notepad. "Those marks Frank makes on his paper are to help him remember things, right Frank?" she says. "I think it's delicious to be so archaic!"

Right about then Perl finds what he's looking for.

The saucer zooms in on a twenty-sixth-century man named Boba Shekk. Boba Shekk is an unkempt fellow with a big nose, prominent wrinkles, thinning hair, and blemished skin. The one attractive thing about him is his broad, slightly deranged smile—not that he smiles very often. As Frank and the saucerians tune in on him, he's engaged in a business call through his uvvy. As usual, the saucerians are translating the future voices into an appropriate-seeming idiom of twentieth-century English.

"I'm working as fast as I can, Manny," Shekk wheedles. "The hurrieder I go, the behinder I get!"

"Boba, Boba, Boba," answers the unseen Manny. "You're contracted to deliver me the pilot for *My Mermaid* by next week. And now I'm hearing you're not outta lifebox devo yet? And, excuse me if I heard you wrong, but you say you want *two more months*? You think Hyena-don Productions is maybe a foundation for the nurture of great artists?''

''My special touch, it takes time,'' snaps Shekk. ''If you want the hurry-up garbage, then go to some schlock crafter who's gonna use off-the-shelf lifeboxes for his simmie actors. But if you want a show that's gonna connect to people's hearts, if you want a hit, then you gotta let Boba Shekk craft you some real characters to carry the plot.''

''Six months of character crafting already?''

''The public only sees one-tenth of the iceberg, Manny. For my characters to sing, they gotta have the life experiences, the quirks, the personal anecdotes. It takes gold to make gold.''

''I'm giving you three more weeks, Boba, and then I'm gonna pull the plug. In fact, I'm gonna call Schneidermann and get him to start work on a backup pilot for *My Mermaid* today. You're not ready in three weeks, we go with Schneidermann. It's your call, big boy.'' Manny closes the connection.

Boba Shekk slumps in the chair niche of his grown home, rubbing his temples. His partner enters the room. She's a Japanese woman with animated green-and-purple hair that echoes her motions with graceful swirls.

''I'm losing it, Etsuko,'' says Boba Shekk. ''The ideas aren't coming. I'm trying to craft as if it doesn't matter, to just put down anything, but even that's not working. I'm gonna lose this gig.''

''Why not get yourself a couple of ohmies to help you?'' says Etsuko. ''We could afford it, just barely. It's a risk, but it could be a powerful investment. I can't stand to see you so unhappy.''

So Boba Shekk makes an uvvy call and orders up two ohmies, biological tank-grown clones of himself.

Balaam skips them forward a week, which is all the time it takes in the twenty-sixth century to force-grow a clone to full adulthood. Frank and the saucerians find Boba Shekk treating the still-unconscious ohmies with his synthesized memory molecules. It's a strange procedure.

Shekk has a red nose and keeps sniffling. Etsuko is watching him from across the round, wooden room. She's wearing a complicated biological filtration mask. The two ohmies, who look exactly like Boba Shekk, are laid out on the floor side-by-side, still unconscious, still wrapped in the pale-blue cloths they came in.

"What if the prions linger?" says Etsuko, her voice muffled from behind the mask.

"They won't," replies Shekk. "The prions are tailored to last no longer than about fifteen minutes. Their gene ends break off extra fast or something." His voice is manic with excitement. He checks the time. "Okay, it's been ten minutes since I infected myself. Meanwhile, the prions have been through thousands of generations and they've started mimicking my memory molecules. All I gotta do to finish programming these handsome ohmies is to sneeze on them. And here it comes."

Boba Shekk leans over the sleeping clones, wildly sneezing. Aboard the saucer, Goola, Perl, and Balaam exclaim in disgust. Meanwhile, the clones continue their slow, deep respiration, breathing in untold numbers of Shekk droplets.

"Your personality is like a disease they're catching?" says Etsuko. "Your memories? What if you went out into a crowd?"

"It would be a beautiful thing, wouldn't it," says Shekk, grinning and loudly blowing his nose. "Gesundheit, baby."

In another five minutes the memory prions have died off and Boba Shekk's nose stops running. But Etsuko's loath to take off her mask.

"You're gonna scare them wearing that thing," says Shekk. "I'm about to wake them up."

"I'm not taking off the mask till tomorrow, Boba," says Etsuko. "If you can handle it, so can your ohmies."

So depleted is Boba Shekk's imagination that he names his ohmies Mark and John. They're a bit soft and unformed-looking, like people who've recently spent time in an institution. But Boba Shekk likes them a lot; he spends the rest of the day kickin' with his ohmies.

The saucer skips forward a few days, and it's evident that the clone-programming process has worked really well. The ohmies are hard-working crafters just like Boba Shekk at his best. They have lots of ideas, but not quite the same ideas as Boba would have had.

"Thought is so chaotic that even the slightest differences in initial conditions send us off on different trajectories," Shekk muses to Mark and John. "It's a wonderful thing. Let's all take a little stroll, each in a

FIGURE 56: Boba Sneezing on His Ohmies

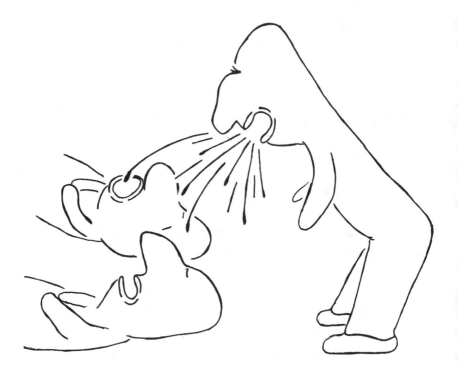

different direction, and get back together with our ideas for *My Mermaid* in an hour or so. I'm sure we'll all come back at the same time anyway.''

The saucer skims along in fast-forward, watching. Boba is totally accepting of what his ohmies do. There's no sense of competition. There's complete unanimity. And there's no financial problems, as Mark and John live with Boba and share all income. The ohmies get their own female partners, Japanese women just like Etsuko, and they add two more house domes to the Shekk compound. Everyone swaps partners freely. It's the perfect idyllic commune.

''But now they'll start fighting,'' says Perl. ''That's the way it always happened with people and their ohmies.''

The saucer jumps a little further, and now Boba is telling Etsuko,

"More and more I feel like there's something fundamentally bogus about the crafting that Mark and John do. They're superficial. Inauthentic. Not real. Not true."

"But Hyenadon likes the new stuff a lot," says Etsuko. "*My Mermaid* is a major hit. Manny's offering you better and better deals."

"I don't care, I want to get rid of the ohmies."

"You're not getting rid of *John*. No way. He's more of a man than you'll ever be."

When Frank and the saucerians check a few months later, they find that John is the kingpin at the Shekk compound. The three women are spending most of their time in his dome, and John rarely uses any of Boba's ideas for the shows he's crafting. As the ultimate indignity, John has started insisting that *he's* the original womb-born Shekk in the group, and that Boba and Mark are his ohmies. Perversely, Mark agrees.

A year later, Boba starts believing it too.

The Ang Ous

"Onward to the fourth millennium," says Perl. "The age of femtotechnology at last."

"I saw some of that on my last trip," Frank says. "I saw about Femtotechnology Unlimited and how they were using big, heavy disks of strange matter to make perfect femtoclones of things. Of people too. And then this woman named, um, Lulu Ma found a way to do it with software or something. It was confusing."

"I'll try and make it simple for you," says Perl. "I'll go on up into the thirty-fifth century, when the alla and femtocloning technology is quite well-established. The copies are completely exact. Eidetic."

"Show him the thousand Ang Ous," suggests Goola. "That was a key turning point."

"Perl's on it, love," says Balaam.

The saucer zeroes in on a building like a honeycomb, down near the waters of the San Francisco Bay. There's a crowd of angry people outside the building. Glaring out of the building's windows are hundreds upon hundreds of identical orange-skinned men, each of the men with the same bowl haircut, each with the same lime-green shirt, each with a thin, wispy mustache. Someone in the crowd yells something unpleasant, and each and every one of the men in the windows shakes his fist. Their raised arms are moving in unison like the bows in the violin section of an orchestra.

"Those are the thousand Ang Ous," says Perl.

"Which one is the real him?" asks Frank.

"They're all equally real," says Perl. "That's the thing about femtocloning. There's no difference, even in principle, between a thing and its femtoclone. Femtocloning specifies an individual's information so exhaustively that the copy is the same as the original. It's just like the way that any two oxygen atoms are the same: Once you say 'eight protons, eight neutrons, and sixteen electrons,' you've said it all. That's how it is with the femtotech description of an Ang Ou."

"But doesn't the original one *know*?" says Frank.

"Frank won't understand till we tune in on Ang himself," says Goola. "Take us closer, Balaam."

The saucer goes into Ang Ou's building, an organically-grown structure with long, thin rooms. The particular room they enter is standing-room-only with Ang Ous. There's even more of them than Frank had seen through the windows.

At the far end of the room is a little titanium tri-bar device with glowing wands of light coming out of it. An alla. The air in the alla box shimmers and a fresh Ang Ou steps out, and then another and another. It's like an automatic bubble-blowing machine that's turning out people.

Frank and the saucerians hop through a series of walls looking at more of the crowded rooms of the Ang Ou building. In each room there is another alla puffing out ever more Ang Ous.

Balaam links Frank into Ang's mind. It's a strange mental space to comprehend, because Ang is seeing out of all of his many bodies' eyes at once. The many brains are linked together by a radiotelepathic contact made perfect by the fact that they're all eidetically the same.

"But they're not wearing uvvies," protests Frank. "How can they do telepathy without uvvies?"

"From the thirty-second century on, people's uvvies are internal," explains Perl. "It's a permanent aug that almost everyone has programmed into their genes. So each of has an electromagnetic uvvy organ in the back of his or her neck. If memory serves, the aug was originally based on some wetware from an electric eel."

As Frank adjusts to the feeling of the Ang consciousness, he begins to feel a sense of exhilaration. So much mental space! It's like having the walls of a confining prison cell fall away to reveal an open plain. What wonders a person could achieve with a thousand brains! Ang's thoughts seem to involve mathematics, physics, and an odd kind of mysticism.

But for now Ang is preoccupied with the hostile street crowd. People are battering down the honeycomb building's doors, forcing their way inside, and fighting to turn off the Ang Ou allas. Sharing in the Ang Ou consciousness, Frank feels like he's struggling with hundreds of people at once. It's terrifying. Ang is fighting like a tiger, careless of which of his bodies lives or dies.

The word of the battle spreads across the city, and more and more people arrive to attack the Ang Ous. Frank is pleased to see that in the fourth millenium, there is no organized police or army, simply the massed will of the populace.

In a short time, the pullulating Ang Ous are overwhelmed. All 1,234 of them are out in the street, tightly bound by their hands and feet. The shared pain of their thousand-plus bodies is vast and somewhat pitiful. The Angs are thinking about pink, sunlit clouds.

"You're so selfish and greedy!" people are yelling into the Ang faces. "You can't stay on Earth!"

The saucer jumps two months further in time, and Frank can see how the Angs have been dealt with. It would have been too inhuman to exterminate them. Instead, they've been subjected to massive amniotic-bath treatments to convert them into leathery spindle-shaped beings with symbiotic colonies of algae in their skins.

"Space bugs," says Goola.

One by one, the Ang Ou space bugs are being launched up through the atmosphere, sent out to live together in space near one of the stable attractor points of the Earth-Moon orbit. A thing something like a cannon is shooting them high out over the Pacific Ocean.

Looking into the distributed mind of Ang Ou, Frank can see that Ang's resigned to this state of affairs. He's planning to do a lot of scientific thinking, out there in space with his thousandfold group mind.

The view through Ang's eyes is remarkable. Some of his bodies are already up in the sky. Merging further into Ang's mind, Frank can get

FIGURE 57: Launching a Space Bug

a slight understanding of the specific problem Ang wants to work on. Ang's thinking about the fact that people need high-tech equipment like allas in order to turn a person into femtotech source-code and to pop the code back into reality. Ang feels like there should be a way to do this solely with the programming of the human mind. A way to make self-extracting personality waves.

Ang feels confident that he will solve the problem. Perhaps it will take his thousand minds two hundred years. That's all right. A space bug should have very good longevity. And when Ang has the answers, the normals will let him return.

TRANSCENDING THE BODY

Teleportation: turn, squeeze, lift. Little Goola like a moth.
The transhumans grow their own saucer.
Eve takes us to the astral plane. It's wonderful. God is love.

Teleportation

"Let's pick up the pace a little," says Balaam, as Perl jumps them back into paratime. "I'm bored, Frank's pooped."

"Don't get bitchy," says Goola. "Not when it's finally time for me to do *my* thing. Would you like me to teach you transcendence, Frank?"

"You bet," says Frank. He's sitting next to Goola, drinking some water out of his eternally refilled cup. "I'm loving every minute of this. I don't want it to be over. I don't want to be back in South Dakota."

"All right," says Goola, rising to her feet. "I'll show you teleportation. It was none other than the Ang Ou swarm who discovered the trick. And in 3666, the Angs taught it to everyone. People were so grateful, they agreed to let the Angs have an Ang Ou island right back here in the San Francisco Bay. They alla-formed it themselves, and there's still Ang Ous living on Ang Ou Island in our own time—in the forty-first century. It's a lush place to visit. The Ang Ous have fabulous gardens."

"Gardens?" wonders Frank. "A superintelligent group-mind like Ang Ou is into *gardening*?"

"Nature is computing ever and always at the maximum possible flop," says Perl. "People never surpass her, they learn only to wonder at her the more. Like me with dear Goola."

"Fast-forward," urges Balaam. "Focus."

"So now I'll show you the day I learned teleportation!" exclaims Goola.

The saucer angles back into time to show a girl sitting in a field of flowers. She's dark-skinned and she has flaxen blond hair. Her eyes are big and alert. The young Goola.

"Oh!" exclaims the Goola in the saucer. "I was so eager then, so tender."

"You still are," says Balaam.

The girl's lips are moving, she's repeating something to herself over and over. And now all of a sudden she disappears and reappears fifty feet away. She does it again and again, hopping back and forth all over the landscape.

Small and invisible, the saucer gets close to the chanting girl. The words she's repeating are surprisingly simple: "Turn squeeze lift. Turn squeeze lift. Turn squeeze lift . . ." Each time she says them, she teleports to a different location.

The saucerians link Frank into the girl's mind, and he can see that the murmured words are connected to specific mental gestures. At each utterance of "turn," the young Goola's view of the world undergoes a kind of mirror-reversal, an inside-out shift. When she says, "squeeze," this altered visual world is somehow condensed into a shining droplet. And with the "lift," the world comes back, but with Goola in a new location.

"You teleport with no technology?" Frank asks the saucerians. "Just by doing some kind of head trick?"

"Oh, there's technology," says Perl. "But it's more like a software than a hardware. It's a script that Ang Ou invented for the human mind. It has to be programmed in. But once you have it—"

Down in the flowery field, some other children have joined Goola. They skitter about the landscape, appearing and disappearing like sudden raindrops on a river. Like fireflies at dusk. Like shooting stars.

Growing a Saucer

"So that's how we learn to flip," says Goola. "Between being solid people and being personality waves."

"And there's more," says Balaam. "Once you can make a body, you can make an object too."

"Such as this flying saucer," says Perl, rapping a knuckle on the metallic wall.

"You were able to mentally create this ship?" asks Frank. "From nothing?"

"I'll show you our takeoff," says Goola, smiling.

The saucer whirls dizzily and now they're into the fifth millennium, hovering high over the region once known as Silicon Valley. The year is 4004. Being drawn so very far into the future gives Frank a hollow feeling in the pit of his stomach.

From the air it looks like pristine wilderness. There's great flocks of birds on the clear waters of the San Francisco Bay, which extends south beyond the old San Jose as far as the eye can see. The hills are covered with oaks, redwoods, and chaparral. The lowlands are green with grasses and scrub. As the saucer drops lower, Frank can see that many of the trees are in fact homes, and that a lot of the chaparral plants bear edible fruit.

The saucer hovers over a ceremonial-looking round clearing in the woods. There's a hissing noise and suddenly one, two, three people appear in the clearing, materializing from nothing but invisible higher-dimensional radiation. It's Goola, Perl, and Balaam. It's strange for Frank to be with them in a saucer watching other versions of them on the ground.

The three figures in the clearing hold hands. A wire-frame outline of a flying saucer appears around them. Curved surfaces fill in the gaps of the wires, slowly solidifying into something like shiny metal.

The metallic saucer spins, vibrates, and shoots up into the sky.

The Divine Reality

"That was such an exciting takeoff," says Balaam. "But—ugh—it makes me sick to watch myself."

"Like looking in the mirror too long," agrees Perl. "Let's finish up with Frank and be on our way out into the cosmos, Goola."

"The cosmos," muses Frank. "Is *everyone* leaving Earth to be a saucerian?"

"Not at all," says Goola. "Many more people are into centering. They dematerialize and rematerialize in the same place. And as they do it, they dip into the astral plane. In fact, that's just about the last thing I'm supposed to show you."

"Eve?" says Perl.

"Yes, now we're going to take Frank to meet Eve," says Goola. "It's always wonderful to see Her."

The saucer goes somewhere in time: perhaps further into the fifth millenium, perhaps back to the dawn of history, perhaps into paratime. Sitting in the shade of a redwood tree by a little stream is a pleasant-looking woman. This is Eve. The saucer hovers before her.

"Hello," says Eve, smiling. It's clear that she can see Frank and the saucerians. Frank feels the calm, gentle touch of her mind.

Eve dips her hand into the stream, then raises her hand up into the sunshine. Her wet fingers sparkle in the light and a rivulet runs from her palm. The flow trickles to a stop, and a last, fat drop of water hangs from the side of Eve's hand.

Something odd happens to the perspective here, and the hanging drop of water looms large and weighty as a moon, as a planet, as a sun. Every fiber of Frank's mind becomes focused upon it.

The immense droplet shimmers, shivers, and then slowly, massively, it pulls free of Eve's hand. Its surface is a great play of complex undulations that seem to be happening ever slower in time.

The sensations generated by the drop's motions extend beyond sight and into touch and sound; Frank can feel and hear them. The touch-sensation of the drop's motions is as the wind of a tropical breeze, like the caressing of a fitful zephyr that patterns the air with whorls and hollows.

The sound-sensations from the drop are like a choir of voices, angelic voices that seem to have been singing forever, voices that Frank's only noticed just now, for the first time in his life, after all these years . . . his eyes fill with tears of joy and wonder.

The thoughts and feelings sent off by the great water drop intensify. Frank feels as if his mind must come unhinged—and perhaps it does.

He's somehow moved *outside of* space, looking into everything from every side. He sees the veins in Eve's body, and the innermost rings of the redwood tree. He sees the worms in the ground and the bubbles in the stream. There are faces and forms to go with the singing voices, elfin bodies that flutter against Frank. Some are aliens, some are human, some are unknown inhabitants of the astral plane.

Eve's calm smile floats behind the great droplet, illuminated by the clear light that shines through everything. And Frank knows that the light is God.

"I'm always here," says God's voice. "I love you, Frank. I'm always here."

Frank breathes in the peace and love, the serenity, the wisdom. He will be a better man.

Eventually, Frank becomes aware of others around him. His eyes flutter, and he's back inside the saucer. Goola, Perl, and Balaam are looking at him.

Frank and Goola

"There's one last thing," says Goola, smiling. She purses her lips and gives Frank's lips a little kiss. She smells so good, or is it really a smell? More like a direct tweaking of the pheromone-sensitive receptors in Frank's nose. Whatever—it works big-time. Frank becomes powerfully aroused.

"Oh, Goola," sighs Frank dizzily.

Goola reaches out her hand and undoes Frank's belt buckle. His pants slide down his legs. Now Goola touches Frank's aching penis, lightly running her fingers along its trembling length. "Why, yes, it might be lifty to take a little wetware sample from you, hmmm? Just like a good saucerian is supposed to do."

"Really, Goola," tut-tuts Perl. "You don't have to behave as if—"

"It's what Frank expects," sighs Balaam. "He needs this for the end of his book. *Mektoub*. But I say we get it over with *toot sweet*."

Goola shucks off her dress, straddles Frank and sits down on his penis, slipping him deep inside her. She kisses Frank full on the mouth.

Outside, the great American West is sweeping by beneath them. Balaam is flying the saucer from San Francisco back to Rapid. The sky is strobing as before, but Frank's too agog to think about it. He trusts the saucerians to do the right thing.

Although the sex act feels very good, it doesn't feel quite—real. Goola's flesh is warm but somehow insubstantial. Slippery, but not wet.

Just as Frank approaches his climax, Balaam maneuvers the saucer back down to the IRC lab, April 9, 1997, and sets Frank down in his office chair. Frank feels the touch of the chair's wood against his bare butt. He ejaculates. There's a tiny tingle near the tip of his penis—Goola decoding a cell's DNA? And then Goola and saucer lift up off of him and bid him the briefest of telepathic farewells.

"Good-bye," murmurs Frank. "Thank you. I love."

The saucerians fade away. Frank is alone in the darkened room with his pants around his ankles, a sticky patch of semen on his shirttail, and his penis slowly softening against his naked thigh. Rather than being wet as if from real sex, the skin of the penis is bone-dry.

And then—whoops!—the door swings open, the lights come on, and there's Helene Lundy staring at Frank, her mouth a tiny O of utter shock and disapproval. It seems that although Helene didn't feel well this morning, she's decided to come in for work this afternoon.

There's a sudden speeded-up rush of screaming and jabbering. The Rapid City police actually get called in, but the investigating officer seems to feel, well—some poor geek of an equipment tech jacking-off in a locked storeroom on his lunch hour—"It's nothing to make a court case out of, Ms. Lundy, although I'm sure it was mighty unpleasant for you. I'm inclined to believe Mr. Shook when he assures me that he would not of been behaving in this manner if he'd of thought there was even the slightest inkling of your walking in on him."

The Black Hills High principal and assistant principal are there to hear this as well. The principal is having trouble keeping a straight face. Helene is such an old maid and Frank such a doofus. This is shaping up to be the funniest mishap of the year. The smiling principal starts to tell Helene to just forget the whole thing and to let Frank get back to his job.

"I bet he's a professional pornographer!" interrupts Helene. "What else would he have been doing with all that TV equipment?"

"I already checked for porno," says the investigating officer. "There's no tapes in the VCRs. Though I suppose Mr. Shook might have one in his coat?"

"No, I don't!" cries Frank, glaring at the bland, mocking faces. "If you must know, I was using the TV equipment to contact a flying saucer. They took me on a trip far away from here. And a saucer woman made love to me."

This doesn't go over well. It's only been a month since the mass suicide of the Heaven's Gate saucer cult members. The amused smiles change to anxious frowns. And Frank the lowly temp worker is fired on the spot.

When Frank finished telling me all this, the sun was setting. We were already making our way down the slope from the Devil's Tower towards our campsite. Although I was incredibly impressed by everything Frank had told me, I felt an impish urge to tease.

"I thought you didn't approve of people who have sex with saucerians," I said, grinning. "Seems like this time you put yourself right on a level with them."

"It wasn't just *sex with Goola*," said Frank. "It was something much higher. It was a mystical vision. And my wetware's going across the galaxies."

"That's what all the abductees say." I laughed.

"You know, Rudy, I'm really sick of you," said Frank, suddenly flaring into extreme anger. "And now, thank God, my part of our book is done and I'll never have to talk to you again. I've been trying to get you to see how rich and wonderful a world we live in, to help you understand a little bit of saucer wisdom, and it's like you're intent on just throwing away everything I tell you."

"I'm sorry, Frank! I take it back! I do understand a little bit. I just couldn't resist—"

"I am so fucking sick of people like you laughing at me."

We didn't talk much more than that. We walked the rest of the way to our camp, ate some noodles, crawled into our tents, and went to sleep.

TWELVE

The End?

That night I had a very unusual experience.

In the middle of the night—this would have been in the early A.M. hours of Monday, June 16, 1997—I woke and looked up through the tent's mesh ceiling to the sky. Directly overhead I saw a brilliant point-sized flash of light that moved along a curving bright line and disappeared—the brightest shooting star I'd ever seen.

Before even forming the thought that I could make a wish, I'd made one. If I were going to lie to you, I'd claim that I wished I could see the aliens. But in fact I wished for something that lies deeper and closer to my heart: I wished to remain sober.

I suppose I went back to sleep then, but a little later I seemed to find myself awake and outside. I was in the high grasses of the field between our campground and the river.

In my dream, an actual metal-just-like-it's-supposed-to-be flying saucer lowered down on the grassy field next to our campground. Time was frozen. The river, which had been babbling before, was still and silent. The headlights of a distant car had stopped moving. In frozen time, the saucer opened a door and drew me in.

Herman the starfish was in there, and Steffi the piece of rope, and some of the iridescent beetles Frank mentioned from before. They were kind and friendly. There were other aliens too: orchids and cacti, crystals and flames, lobsters and lizards, clouds and whirlwinds, all of them chirping and smiling and crowding forward.

Oddly calm, I asked how they all came to be in the same saucer. Frank had always seen only one kind of alien at a time.

"We're from even further out in the dimensions," a sea cucumber named Tekel told me. "We're from the time beyond paratime. Where everyone ends up."

"Are there any humans aboard?" I didn't see any, but I was kind of hoping to get a look at Goola.

Reading my thoughts, a beet named Gordo gave a cryptic answer, "She's not here yet. Not yet yet. Not yet yet yet. Three more levels."

The aliens told me to be sure and finish *Saucer Wisdom*. To not be scared. That they'd watch over me, to the extent that they could. "And just trust God," was the last thing the aliens told me. "God will protect you. God's always there."

When I woke up in the morning, Frank Shook was gone. No trace of him remained—and strangest of all, the tent of his I'd been sleeping in was gone as well. I was lying there on the bare ground of our campsite in my sleeping bag, with my glasses, clothes and water bottle on the dirt at my side. The money from my wallet, some three hundred dollars, was gone. Oh well. The vision had been worth it.

I wandered down towards the river, into the field where my dream had taken place. I felt an odd lack of surprise to find a spot where the grasses were flattened in an approximately circular pattern, whirled around as if by a small tornado. Black crickets hopped everywhere underfoot.

I packed up and got in my rented car. My plane home wasn't till late that afternoon, so I went by a Safeway to get some food and cash with my credit card.

In the Safeway, my thoughts turned to the superfluity of the endless Midwestern Safeways by the interstates, the Safeways and their identically-polite checker girls with Norwegian *oooo* sounds in their speech. And the superfluity of the identical old farmers and farmwives, spry and cantankerous, making the best of things. The endless glut of these people repeated over and over with the same expressions and opinions, they seemed to me like a field of flowers. All the same. In a way this bothered me, in a way it didn't. Why not repeat, after all, that's how fields of flowers are, all the same, and it's just a Romantic error to expect the windrows of humanity to be anything other than fields of people, the same pattern duplicated and reduplicated. Nature likes repeating herself.

And as I thought of the reduplicating flower-flesh of humanity, I could fully see how deep a disinterest the aliens have in us. The aliens have as deep a don't-care feeling about us as my feelings upon seeing yet another squirming black cricket in the river field where I walked and wondered that Monday morning after my dream of the saucer from beyond time.

I haven't heard any more from Frank Shook. I have no idea where—or when—he is. Perhaps Finland? I have a feeling he'll turn up again some day, so I'm saving his share of the *Saucer Wisdom* money—minus three hundred dollars—in a special account. My agent and I made several attempts to look up Peggy Sung, but we could only learn that she'd moved from Benton to either Orange County or to mainland China. And as for Frank's ex-wife Mary, the phone number she gave me doesn't work anymore, I don't know her last name, and nobody in San Lorenzo seems to have any idea how to find her.

Some friends and I pooled our video cameras and TVs and tried to reproduce Frank's alien-attracting technique, but we couldn't get it to work. We saw a lot of odd images, but that was about as far as it got us.

In any case, I've come to like thinking about aliens; I sometimes even imagine that I really did spend a few minutes in a saucer that night by Devil's Tower. It's a fresh, spaced-out way to look at the world. A conceptual high. I often think of a UFO perched watching at my shoulder, and it makes me feel glad.

And no, I haven't been stopped from telling this True Story, and you reading this, no, you aren't letting anyone stop you either, you're in on the secret now, you're in the Big Time, you've learned Saucer Wisdom.

The aliens are all around us, and you can learn to see things as they do.

God is everywhere, and if you ask, God will help you.

Wisdom enough.

INDEX

ABC of Relativity, The (Russell), 34
Abduction: Human Encounters with Aliens (Mack), 21, 24
Abbot
   and real-time programmable displays, 72
Abbot wafers, 72
aliens
   appearance of, 108–13
   Blukka (alien spine-riders), 131
   brain etching, 62–64
   life in the universe, 130–38
   number of visits from, 130
   robots, opposition to, 142–43
   rope aliens, 208–13
   shuggoth world, 133–34
   sunspots as a life form, 134–35
   travel as a higher-dimensional electromagnetic wave, 116–17
   why they're here, 113–15
aliens (notes on), 105–38
   cosmic road trip, 106–15
   life in the universe, 130–38
   personality waves, 115–30
alla (immersive and interactive virtual reality), 213–17
   and the asteroid belt, emigration to, 222–26
   expense and success of, 220–21
   Femtotechnology Unlimited and, 220–21
alla wands, 120–23
Alpha Centauri
   emigration to, 226
animals (chemical beasts), 142–58
   Big Tongue, 143–44
   Biobot Inc., 143–44
   FlushFish, 152–54
   Neatnick Birds, 152–54
   pet dinosaurs, 155–58
   Phido (pet-construction kit), 154–58
   robots, opposition to, 142–43
   ShushBees (reengineered bumblebee), 147–50
   ShushBees, application of, 150–52

wetware engineering, 147–50
   See also humans; plants
apps
   fabulous (wigged in), 83–84
asteroid belt
   emigration to, 222–24

Balaam (androgyne saucerian)
   and Aug Dogs (teenagers), 245–48
   and Ang Ous multiples, 265–69
   and the archipelago people, 248–49
   copying the body, 256–69
   divine reality, 271–73
   Frank and, 240–45
   Frank's sex with Goola, 273–75
   growing a saucer, 270–71
   and mermen, 253–56
   and the NuYu Do-It-Yourself Fourteen-day Rejuvenation bath, 250–53
   Boba Shekk's ohmies, 261–65
   Mr. Sloane's clone, 257–61
   teleportation, 269–70
Bark, Carl, 86
bat man, 170
   See also morphed humans
"Beast of Burden" (Rolling Stones song), 205
Big Tongue, 143–44
   vs. Lickin' Luv Slug, 145–47
bigwings (mind modem)
   and fabulous apps, 83–84
   transmitted emotions, 82–83
   uvvies, 84, 84–85
Biobot Inc. (San Mateo), 143–44
   Big Tongue, 144–44
   Big Tongue vs. Lickin' Luv Slug, 145–47
   FlushFish, 152–54
   Neatnick Birds, 152–54
   ShushBees (reengineered bumblebee), 147–50
   ShushBees applications, 150–52
   wetware engineering, 147–50
biotechnology (note on future), 139–72
   chemical beasts, 142–58

biotechnology (*continued*)
  morphed humans, 170–72
  tweaked plants, 159–69
Blukka (alien spine-riders), 131
body copying, 256–69
  the Ang Ous multiples, 265–69
  Boba Shekk's ohmies, 261–65
  Mr. Sloane's clone, 257–61
body editing, 240–56
  Aug Dogs (teenagers), 245–48
  and the archipelago people, 248–49
  extra hands, 249, 250–51
  flying eyeball, 253, 253–54
  mermen, 253–56
  NuYu Do-It-Yourself Fourteen-day
    Rejuvenation bath, 250–53
body transcending, 269–75
  divine reality, 271–73
  growing a saucer, 270–71
  teleportation, 269–70
Book of Numbers (Conway/Guy), 119
bumblebees (reengineered)
  ShushBees, 147–50
  ShushBees applications, 150–52
brain concert
  Larky's, 78–79
brain etching, 62–64

Carlo (Milla Maize's lover), 61–62
Carol
  and cloning Mr. Sloane, 258–61
  and the S-cube, 258–61
Chad (chemical engineer), 70–71
*Chaos and Fractals: New Frontiers of
  Science* (Peitgen/Jurgens/Saupe), 47–
  48n.8
chaotic (limpware) engineering, 69–70
chemical beasts, 142–58
  Big Tongue, 143–44
  Biobot Inc., 143–44
  FlushFish, 152–54
  Neatnick Birds, 152–54
  pet dinosaurs, 155–58
  Phido (pet-construction kit), 154–58
  robots, opposition to, 142–43
  ShushBees (reengineered bumblebee),
    147–50
  ShushBees applications, 150–52
  wetware engineering, 147–50
  See also morphed humans; tweaked
    plants

Chu (kid)
  using Phido (pet-construction kit) to
    engineer dinosaurs, 155–58
cloning
  ohmies, 261–65
  Ang Ous multiples, 265–69
  Mr. Sloane, 257–61
Close Encounters of the Third Kind
  (film), 204, 239
communications (notes on the future of),
  54–89
  dragonfly cameras, 59–64
  lifebox, 55–59
  piezoplastic, 64–76
  radiotelepathy, 76–89
Communion: A True Story (Strieber), 21,
  22
Conrad
  and bigwigs, 84
context
  lifebox database, 59
cosmic road trip, 106–15
  the aliens, 108–13
  purpose of visit, 113–15
  the saucer, 106–8
cosmic snow
  and video feedback, 45–50
Conway, John H., 119

Daisy Hill Puppy Farm (Phido-
  compatible incubatorium), 154–55
  using Phido (pet-construction kit) to
    engineer dog, 154–55
Dak
  and soul broadcast, 117–19
Dekker, Desmond, 93
Devilberries, 163
Devil's Tower, 203–6, 239–40
  trails, 207
divine reality, 271–73
Dieu
  and cloning Mr. Sloane, 258–61
dinosaurs (pet)
  using Phido (pet-construction kit) to
    engineer, 155–58
Dogon (Zen monk), 128
Don (sheriff deputy)
  and the burglary investigation, 188–89
Dona (kid)
  using Phido (pet-construction kit) to
    engineer dinosaurs, 155–58

dragonfly cameras, 59–64
  alien brain etching, 62–64
  dragonfly news, 59–61
  dragonfly paparazzo, 61
  gnat cameras, 63
  renting, 63
Drake, Frank, 46
dreamtime
  and paratime, 32
Duckburg (Bark), 86
Durleen (teenager with finger necklace),
  245
  and the High Plains Smoothie, 245–48
Dustin (the snail man), 67
  and the appearance of plastic slugs (3–
  D paisley), 67–69

Eckhart, Meister, 32
Einstein, Albert, 38
Embrey, Camilla (wife)
  flying eyeball, 253, 253–54
  and the NuYu Do-It-Yourself Fourteen-
  day Rejuvenation bath, 250–53
Embrey, Otis (farmer)
  extra hands, 249, 250–51
  and the NuYu Do-It-Yourself Fourteen-
  day Rejuvenation bath, 250–53
  and pharmaceutical cows, 249
Embrey, Rize (son)
  as a merman, 253–56
Etsuko
  using ohmies, 261–65
Eve
  Frank meeting, 271–73
extra hands
  Ottis Embrey and, 249, 250–51

Femtotechnology (notes on), 207–38
  asteroid belt, emigration to, 222–26
  matter transmutation, 208–26
  and space travel, 222–26
  3oxing, 226–38
Femtotechnology Unlimited, 217–20
  alla, expense and success of, 220–21
  cloning collectibles, 231–32
  and Lula Ma's alla-ended shuffleboard
  stick, 235–37
  matter maps, 232–37
and strange matter, 226–31
Floto (hiker)
  using the alla, 213–17

FlushFish, 152–54
flying eyeball, 253–54
  and the NuYu Do-It-Yourself Fourteen-
  day Rejuvenation bath, 250–53
Foo-Foo (LuvSlug), 65–66
four breasted woman, 170
  See also morphed humans
Fourth Dimension, The (Rucker), 22, 34
Freeware (Rucker), 194–95
fury fish eating arctic couple, 170
  See also morphed humans

Gandy, Dick (Frank's neighbor), 41, 201
Gandy, Hank (Frank's neighbor), 41,
  201
Gandy, Sharon (Frank's neighbor), 41,
  201
  and Frank and Mary's departure, 193–
  94
Gay, Benny
  and the contest between Big Tongue &
  Lickin' Luv Slug, 145–47
Gene (Biobot engineer)
  and FlushFish, 152–54
  and Neatnick Birds, 152–54
  and ShushBees (reengineered
  bumblebee), 147–50
  ShushBees applications, 150–52
Giant's Beanstalk, Inc. (biotechnology
  company)
  and Devilberries, 163
  and the Giga Gourd, 163–64
  and King Kong carrots, 163
  and knifeplants, 159–62
  and Pontoon peppers, 163
  and Supergarlic, 162–63
gibbon armed woman, 170
  See also morphed humans
Giga Gourds, 163–64
gnat cameras, 63
God
  encounters with, 136–37, 273
Goola (female human saucerian)
  and the archipelago people, 248–49
  and Aug Dogs (teenagers), 245–48
  copying the body, 256–69
  divine reality, 271–73
  Frank and, 240–45
  Frank's sex with, 273–75
  growing a saucer, 270–71
  and mermen, 253–56

Goola (continued)
    and the NuYu Do-It-Yourself Fourteen-
        day Rejuvenation bath, 250–53
    Ang Ous multiples, 265–69
    Boba Shekk's ohmies, 261–65
    Mr. Sloane's clone, 257–61
    teleportation, 269–70
GRBs (gamma ray bursts), 180
Green Balls, 167–69
Grown Homes
    Green Balls, 167–69
    Heironymous Bosch House, 165–66
Guster (Spun's friend)
    and the burglary, 201, 209
    Mondo 2000 party, 181–83
    selling Lotus Lights, 182–83
Guster (Frank's friend), 97–98
Gutierrez, Jose and Amparo (Gilroy
        farmers)
    and Devilberries, 163
    and the Giga Gourd, 163–64
    and King Kong carrots, 163
    and the knifeplants, 159–62
    and Pontoon peppers, 163
    and Supergarlic, 162–63
Guy, Richard K., 119

Harry (Femtotechnology Unlimited
        founder)
    and Femtotechnology, 217–20
    and Lula Ma's alla-ended shuffleboard
        stick, 235–37
    and nanotechnology, 217–18
    and strange matter, 226–31
Henry and Diana (vacationing couple)
    and Green Balls, 167
Herman (an alien), 276
    appearance, 108–13
    and chemical beasts, 142–58
    creating a saucer, 126–28
    home planet of, 123–25
    and life in the universe, 130–38
    and matter rays, 120–23
    robots, opposition to, 142–43
    the smell of souls, 125
    and soul broadcast, 117–19
    travel as a higher-dimensional
        electromagnetic wave, 116–17
    why they're here, 113–15
Hermit Huck (Gerard Walloon figurine),
    231

Hogben, Lancelot, 34
housing
    Green Balls as, 167–69
    Grown Homes and, 164–66
humans (morphed humans), 170–72
    bat man, 170
    four breasted woman, 170
    fury arctic couple, 170
    man with kangaroo tail, 170
    new born child with augmented (625
        finger tips) left hand, 170
    stegosaurus man, 170–71
    tough-skinned spindle pod person, 172
    truck driver with wimpled roll shaped
        head, 172
    woman with gibbon arms, 170
    See also animals; plants
Huxley, Aldous, 27

Introduction to Zen Buddhism (Suzuki),
    27
Is Anyone Out There? (Drake/Sobel), 46

Jean (Biobot engineer)
    and FlushFish, 152–54
    and Neatnick Birds, 152–54
    and ShushBees (reengineered
        bumblebee), 147–50
    ShushBees applications, 150–52
Jean (teenager with multiple body parts)
    and the High Plains Smoothie, 247–
        48
Jeremy (cross-dressing dragonfly camera
        person), 61
Long Tooth Noser Girl (UV show), 62
Jerry (slug hacker)
    3-D cellular automata (CA), 70
Jimi (child)
    using Phido (pet-construction kit) to
        engineer dog, 154–55
Joe (Femtotechnology Unlimited
        employee)
    alla improvements, 220–21
    and Femtotechnology, 217–20
    and Lula Ma's alla-ended shuffleboard
        stick, 235–37
    and nanotechnology, 217–18
    and strange matter, 226–31
John (Shekk's ohmic), 264–65
Jung, Carl, 18, 19
    on sexual instincts, 20

kangaroo tailed man, 170
   See also morphed humans
King Kong carrots, 163
knifeplants, 159–62
Kotona (superanimator), 86

Larky (performer), 78–79
   brain concert, 78–79
   and telepathy feedback, 79–80
   and transmitted emotions (mind
      modem), 82
Lasser, Uli (architect-developer)
   and Grown Homes, 164–66
life in the universe, 130–38
   alien guest book, 130–31
   God, 136–37
   new style of urology, 138
   sunspots as a life form, 134–35
   temporal attractor, 137–38
   wetware worlds and otherwise, 131–34
lifebox, 55–59
   cities like lichen, 56
   context (database), 59
   grandpa Ned and, 57–59
   "me-shows," 56
lightspeed travel, 128–30
limpware (chaotic) engineering, 69–70
Lona (collectible dealer)
   cloning collectibles, 231–32
   and 3oxing, 232–33
Lotus Lights, 182–83
Louis (sheriff deputy)
   and the burglary investigation, 188–
      89
Lucy (Larky's lover), 79–80
   and transmitted emotions (mind
      modem), 82
Lula Ma (scientist)
   and matter rays (alla wand), 120–23
Lundy, Helene
   and Frank's display, 273–75
LuvSlug, 65–66
   vs. Big Tongue, 145–47

Ma, Lula
   and the alla-ended shuffleboard stick,
      235–37
   and the Hilbert space spectrum, 235
Mack, John, 21, 24
Maize, Milla (pop superstar), 60–61, 61–
      62

mandalas, 19–20
Manny
   and the My Mermaid pilot, 261–65
Mark (Shekk's ohmic), 264–65
Mathematics For the Million (Hogben),
      34
matter maps, 232–37
   yottahertz Hilbert chirp and, 235–36
matter transmutation, 208–26
   the alla, 213–17
   rope aliens, 208–13
mind modem (bigwings)
   and fabulous apps, 83–84
   transmitted emotions, 82–83
   uvvies, 84, 84–85
Mondo 2000, 173–78
   party, 178–83
Mondo 2000: A User's Guide (Rucker/
      Sirius/Mu, eds.), 174
morphed humans, 170–72
   bat man, 170
   four breasted woman, 170
   fury fish eating arctic couple, 170
   man with kangaroo tail, 170
   new born child with augmented (625
      finger tips) left hand, 170
   stegosaurus man, 170–71
   tough-skinned spindle pod person (in
      space), 172
   truck driver with wimpled roll shaped
      head, 172
   woman with gibbon arms, 170
   See also chemical beasts; tweaked
      plants
Mount Rushmore, 233–35, 237–38
Mu, Queen, 174
   Mondo 2000 party, 178–83

nanotechnology, 217–18
Natur, Kenny, 175–76, 201
Neatnick Birds, 152–54
Ned, Grandpa
   and the lifebox, 57–59
Nelda
   and S-cubes, 87–89
new born child with augmented (625
      finger tips) left hand, 170
   See also morphed humans
Newk (skater-engineer), 66
NuYu Do-It-Yourself Fourteen-day
      Rejuve-nation bath, 250–53

Omid (limpware engineer)
  uvvy files and superanimation, 85–86
"Once Bitten Twice Shy" (song), 205
Ous, Ang
  multiples of, 265–69

Pascal, Blaise, 19
Pepita, Señor
  and Devilberries, 163
  and the Giga Gourd, 163–64
  and King Kong carrots, 163
  and the knifeplants, 160
  and Pontoon peppers, 163
  and Supergarlic, 162–63
Perennial Philosophy (Huxley), 27
Perl (male human saucerian)
  and the archipelago people, 248–49
  and Aug Dogs (teenagers), 245–48
  copying the body, 256–69
  divine reality, 271–73
  Frank and, 240–45
  Frank's sex with Goola, 273–75
  growing a saucer, 270–71
  and mermen, 253–56
  and the NuYu Do-It-Yourself Fourteen-
    day Rejuvenation bath, 250–53
  Ang Ous multiples, 265–69
  Boba Shekk's ohmies, 261–65
  Mr. Sloane's clone, 257–61
  teleportation, 269–70
personality waves, 115–30
  creating a saucer, 126–28
  decrypting, 119
  lightspeed travel, 128–30
  matter rays, 120–23
  the smell of souls, 125
  and soul broadcast, 117–19
  travel as a higher-dimensional
    electromagnetic wave, 116–17
pet dinosaurs
  using Phido (pet-construction kit) to
    engineer, 155–58
pharmaceutical cows, 249
Phido (pet-construction kit) using to
    engineer dinosaurs, 155–58 using to
    engineer dog, 154–55
piezoplastic, 64–76
  appearance (3-D paisley), 67–69
  chaotic (limpware) engineering, 69–70
  Dustin the snail man, 67, 67–
    69

LuvSlug, 65–66
Newk's Oaktown slugskates, 66
polyglass, 76
sewer slugs, 64–65
sluggie processors, 73–74
smart furniture 74–76
soft (real-time programmable) displays,
    72
Texas machine language, 70–71
plants (tweaked plants), 159–69
  Devilberries, 163
  Giga Gourd, 163–64
  Green Balls, 167–69
  Grown Homes, 164–66
  King Kong carrots, 163
  knifeplants, 159–62
  Pontoon peppers, 163
  Supergarlic, 162–63
  See also animals; humans; plants
Pol (dweeb collector)
  cloning collectibles, 231–32
  and 3oxing, 232–33
polyglass, 76
Pontoon peppers, 163
pterandons
  as a pet, 158

quantum dot energy source, 2221
Quantum Reality, 34
Quarkaese, Suzanna
  and the contest between Big Tongue
    & Lickin' Luv Slug, 145–47
quark-flipping, 219

radiotelepathy, 76–89
  fabulous apps (wigged in), 83–84
  Larky's brain concert, 78–79
  life in a saucer, 77
  the mind modem, 82–83
  recording dreams, 87–89
  soft satellites, 84–85
  telepathy feedback, 79–80
  uvvy files and superanimation, 85–
    86
recording dreams
  S-cubes and, 87–89
rhamphorhynchii
  as a pet, 158
Richards, Keith, 205
Robinson, Kip (Santa Cruz hotelier)
  and Green Balls, 167–69

Robinson family (Ali, Ala, Ben, Bea, Cus
   and Cis)
   asteroid belt, emigration to, 222–26
robots
   opposition to, 142–43
   See also animals; humans; plants
rope aliens, 208–13
Rucker, Audrey
   and the burglary, 184–87
   and the burglary investigation, 188,
      190
   jellyfish paintings, 53, 188, 194
   Mondo 2000 party, 173–83
   and Frank Shook, 25, 99–100
Rucker, Rudy
   novels of, 194–95
   prayer, 26–27
   as a reviewer, 24
   on ufology, 18–22
   and UFOs, 17–18
Rucker, Rudy (and Frank Shook)
   biotechnology notes, 139–42
   and the burglary, 184–87
   and the burglary investigation, 188–
      94
   communications, notes on the future
      of, 54–89
   cosmic snow and video feedback, 45–
      50
   Frank's house, 41–42
   Frank's house, driving to, 39–40
   Mondo 2000 party, 173–83
   notes on aliens, 105–38
   notes for Saucer Wisdom, 90–100
   saucer demo, 50–53
   Bruce Sterling on, 14–16
   transhumanity, notes on, 239–75
Rucker, Rudy (and Frank Shook:
   meetings)
   Devil's Tower, 203–6, 207, 239–40
   first meeting (Dominican College
      lecture), 22–26
   Jahva House, 90–100
   San Lorenzo, 28–30
   South Dakota, 197–203
Russell, Bertrand, 34
Rust (hiker)
   using the alla, 213–17

S-cubes, 86
   Nelda and, 87–89

Saleem (Biobot engineer)
   Big Tongue, 143–44
   ShushBees, application of, 150–52
Sara (child)
   using Phido (pet-construction kit) to
      engineer dog, 154–55
saucer demo, 50–53
Saucer Wisdom (Rucker), 137–38, 196–97
   burglary and, 184–87
   notes, deciphering, 90–100
   notes on aliens, 105–38
   old copy of (in the future), 118
   Peggy Sung's criticism of, 208–9
saucers, 106–8
   creating a saucer, 126–28
   life in, 77
   sewer slugs, 64–65
Shekk, Boba (creative talent)
   and the My Mermaid pilot, 261–65
   ohmies, using, 261–65
Shirley (slug hacker)
   and the appearance of plastic slugs
      3-D paisley), 67–69
Shook, Frank
   Mary, separation from, 240–41
   and paratime, 30–39
   and rope aliens, 208–13
   sex with Goola, 273–75
   and Peggy Snug, 100–104
   Bruce Sterling on, 14–15
Shook, Frank (and Rudy Rucker)
   biotechnology, notes on future, 139–72
   burglary, 184–87
   communications, notes on the future
      of, 54–89
   cosmic snow and video feedback, 45–
      50
   Femtotechnology, notes on, 207–38
   Frank's house, 41–44
   Frank's house, driving Rudy to, 39–40
   and the human saucerians, 240–45
   Mondo 2000 party, 173–83
   and Mount Rushmore, 233–35, 237–38
   notes on aliens, 105–38
   saucer demo, 50–53
   transhumanity, notes on, 239–75
Shook, Frank (and Rudy Rucker:
   meetings)
   Devil's Tower, 203–6, 207, 239–40
   first meeting (Dominican College
      lecture), 22–26

Shook, Frank (*continued*)
  Jahva House, 90–100, 137–38
  San Lorenzo, 28–30
  South Dakota, 197–203
Shook, Mary, 31, 42–43, 51–6, 278 2
  and the burglary investigation, 188–89
  Frank, separation from, 240–41
  Mondo 2000 party, 173–83
  and the South Dakota meeting, 197
shuggoth world, 133–34
ShushBees (reengineered bumblebee), 147–50
  application of, 150–52
Sirius, R. U., 174
  Mondo 2000 party, 178–83
Sloane, Mr.
  cloning, 257–61
slugs (piezoplastic)
  appearance (3–D paisley), 67–69
  LuvSlug, 65–66
  sewer slugs, 64–65
  sluggie processors, 73–74
  slugskates, 66
  smart furniture 74–76
slugskates, 66
smart furniture 74–76
Sobel, Dava, 46n.7
soft (real-time programmable) displays, 72
soft satellites uvvies and, 84–85
Software (Rucker), 194
space bugs, 267–69
Spun (Frank's friend), 96–97, 176
  and the burglary, 201, 209
  Mondo 2000 party, 181–83
  selling Lotus Lights, 182–83
Stapledon, Olaf, 14
Steffi (rope alien), 212–13, 276
  and the alla, 213–17
  and Femtotechnology Unlimited, 217–20
  returning Frank to Mount Rushmore, 233–35, 237–38
  and 3oxing, 232–33
stegosaurus spined man, 170–71
  See also morphed humans
Sterling, Bruce
  on Frank Shook, 14–15
Stig (teenager with lizard tail), 245–46
  and the High Plains Smoothie, 245–48
strange matter, 226–31

Strieber, Whitley, 21, 22
sunspots
  as a life form, 134–35
Sung, Peggy, 43–44, 278
  and the burglary investigation, 190–94
  and flying saucers, 176
  Frank and, 100–104
  and rope aliens, 208–13
  Saucer Wisdom, criticism of, 208–9
superanimation
  uvvy files and, 85–86
Supergarlic, 162–63
Suzuki, D. T., 27
Szilvia (Omid's lover), 86

Tekel (sea cucumber), 277
Telepath, Inc.
  and transmitted emotions (mind modem), 82
telepathy feedback, 79–80
temporal attractor, 137–38
Texas machine language, 70–71
Thieu
  and cloning Mr. Sloane, 258–61
3-D cellular automata (CA), 70
3-D paisley plastic slugs, 67–69
3ox disk, 226–38
  cloning collectibles, 231–32
  strange matter, 226–31
  and the wealthy old man, 233
tough-skinned spindle pod person (in space), 172
  See also morphed humans
transcending the body, 269–75
  divine reality, 271–73
  growing a saucer, 270–71
  teleportation, 269–70
transhumanity (notes on), 239–75
  copying the body, 256–69
  editing the body, 240–56
  transcending the body, 269–75
travel
  as a higher-dimensional electromagnetic wave, 116–17
  lightspeed travel, 128–30
Turla
  and decrypting personality waves, 119
  and soul broadcast, 117–19
tweaked plants, 159–69
  Devilberries, 163
  Giga Gourd, 163–64

Green Balls, 167–69
Grown Homes, 164–66
King Kong carrots, 163
knifeplants, 159–62
Pontoon peppers, 163
Supergarlic, 162–63
See also chemical beasts; morphed
    humans
ufology, 18–22
conspiracy theories and, 22
new style of, 138
obsession with political power, 21–22
role of sex in, 21, 22
UFOs, 17–18
life in a saucer, 77
universe, life in the, 130–38
alien guest book, 130–31
God, 136–37
new style of urology, 138
sunspots as a life form, 134–35
temporal attractor, 137–38
wetware worlds and otherwise, 131–34
UV (universal viewer), 56
and fabulous apps (wigged in), 83–84
uvvies, 84
sort satellites and, 84–85
uvvy files and superanimation, 85–86

video feedback
cosmic snow and, 45–50
experiments on, 47–48n.8
Volga (older Russian man)
and the High Plains Smoothie, 247–48

Wes (titular editor of Mondo), 178
Weston, Ace
and bigwigs, 84
Wetware (Rucker), 194
wetware engineering, 147–50
FlushFish, 152–54
Neatnick Birds, 152–54
and ShushBees (reengineered
    bumblebee), 147–50
ShushBees applications, 150–52
wetware worlds, 131–34
definition of, 131
non-wetware world, 133–34
number of alien visitors to, 130
wimpled roll shape headed truck driver,
    172
See also morphed humans
Wubbo base (South Pole of the Moon),
    221
Wyla (kid)
using Phido (pet-construction kit) to
    engineer dinosaurs, 155–58

Xerxes
Robinson's emigration to, 222–26
X-Files (TV series), 19

Yanno (Lula Ma's husband)
and matter rays (alla wand), 120–
    23

zero-EQ, 95
zettabytes, 119
zodiac as a mandala, 19–20